SpringerBriefs in Environmental Science

SpringerBriefs in Environmental Science present concise summaries of cutting-edge research and practical applications across a wide spectrum of environmental fields, with fast turnaround time to publication. Featuring compact volumes of 50 to 125 pages, the series covers a range of content from professional to academic. Monographs of new material are considered for the SpringerBriefs in Environmental Science series.

Typical topics might include: a timely report of state-of-the-art analytical techniques, a bridge between new research results, as published in journal articles and a contextual literature review, a snapshot of a hot or emerging topic, an in-depth case study or technical example, a presentation of core concepts that students must understand in order to make independent contributions, best practices or protocols to be followed, a series of short case studies/debates highlighting a specific angle.

SpringerBriefs in Environmental Science allow authors to present their ideas and readers to absorb them with minimal time investment. Both solicited and unsolicited manuscripts are considered for publication.

More information about this series at http://www.springer.com/series/8868

Gunther Schmidt • Simon Schönrock
Winfried Schröder

Plant Phenology as a Biomonitor for Climate Change in Germany

A Modelling and Mapping Approach

 Springer

Gunther Schmidt
University of Vechta
Vechta, Germany

Simon Schönrock
University of Vechta
Vechta, Germany

Winfried Schröder
Chair of Landscape Ecology
University of Vechta
Vechta, Germany

ISSN 2191-5547   ISSN 2191-5555 (electronic)
ISBN 978-3-319-09089-4   ISBN 978-3-319-09090-0 (e-Book)
DOI 10.1007/978-3-319-09090-0
Springer Cham Heidelberg New York Dordrecht London

Library of Congress Control Number: 2014944561

Printed on acid-free paper

Springer is part of Springer Science+Business Media (www.springer.com)

Contents

List of Figures

List of Tables

Chapter 1
Background and Goals

Abstract The investigations presented refer to the development of plant phenology since the 1960s in Germany (*case study 1*) and in one of its 16 federals states, Hesse, located in central Germany (*case study 2*). Whereas *case study 1* concentrated on the development of six plant phases in an countrywide overview for whole Germany, *case study 2* deepened the knowledge by example of the plant phenological development of 35 phases in Hesse depicting the whole growing season and using more accurate data on air temperature development in the past and by applying projected air temperature data considering four instead of two currently discussed climate models. In Hesse, the average air temperature of the period 1991–2009 was by 0.9 °C higher compared to the climate reference period 1961–1990. The same holds true for entire Germany, whilst in some regions the temperature increase was even more than 3 °C. According to the respective climate projection, air temperatures in Germany and Hesse, in particular, are expected to rise by about 3.7 °C in maximum until the end of the twenty-first century compared to the reference period 1971–2000. The already observed and even accelerating future air temperature rise may affect the beginning and the length of phenological stages (phases) of plants causing ecological and economic impacts. In order to deal with this issue, the two case studies presented should analyse and model spatiotemporal trends of plant phenology as being an indicator for climate change related biological effects. The phenological data on wild growing plants, fruits, crops and vines were compiled in a GIS indicating different phenological seasons. The data on plant phenology were collected at more than 6500 sites in Germany (553 in Hesse) between 1961 and 2009. Estimations on phenological development in future periods were based on data from two (*case study 1*) and four (Hesse) regional climate models, respectively (*case study 2*). The statistical association between air temperatures and onset of phenological phases for past climate periods was investigated by regression analysis. Regression models derived from statistical analysis of data on the past development of plants were applied on projected temperature maps to estimate future phenological trends. For phases showing a significant and at least medium correlation ($|r| \geq 0.5$), phenological maps referring to past and future periods were calculated by Regression Kriging. Both case studies revealed that a distinct shift of phenological onset towards the beginning of the year is already detectable when comparing the past periods 1961–1990 and 1991–2009. For Germany, the strongest shift was observed for hazel bloom advancing 13 days to the beginning of the year. In future, a shift of up to 33 days was calculated comparing data of 1961–1990 and

G. Schmidt et al., *Plant Phenology as a Biomonitor for Climate Change in Germany*, SpringerBriefs in Environmental Science, DOI 10.1007/978-3-319-09090-0_1

1

2051–2080. In Hesse, 31 out of 35 phases started earlier in the years 1991–2009 compared with 1961–1990. These shifts were stronger in Hesse (8 days in average) than in Germany (6 days in average). Several phases even shifted up to more than 10 days in average. As winter phases tend to shift towards the end of the year, this yielded a prolongation of the vegetation period amounting up to 3 weeks. In Hesse, more than 70% of the phases in each of the past periods were correlated with air temperature by $r \leq -0.5$, more than 50% even by $r \leq -0.7$. Since the 1990s, phenological shifts and regional differences in phase onsets amplified. In many cases, shifts between 2071–2100 and 1961–1990 are expected to be at least twice as high as between 1991–2009 and 1961–1990. The findings on the statistical relation between temperature rise and phenology development have relevance for development of mitigation measures considering environmental, agricultural and economic issues emerging from changes in plant growth and distribution due to climate change.

Keywords Biomonitoring · Climate change impacts · Geostatistics · Plant phenology · Regression Kriging

Ecological impacts of climate change, occurring on several levels of environmental systems' hierarchy, are broad and diverse. They include, for instance, altering species' distribution ranges, plant phenology and growth, carbon and nutrient cycling (Schröder and Pesch 2011), as well as biodiversity and extinction risk (Foden et al. 2013). Significant effects on the seasonal timing of developmental stages could even be corroborated for species regulating their body temperature within a wide range of ambient temperatures (Caro et al. 2013; Polgar et al. 2013). Insects such as mosquitos may profit from higher temperatures and evapotranspiration, and—since they may act as vectors for pathogens—the spread of diseases may be enhanced (Schröder and Schmidt 2014). Mouillot et al. (2013) could indicate the importance even of rare species preservation since rare species disproportionately increase the potential breadth of functions of ecosystems and, thus, are likely to ensure from future uncertainty arising from climate change. Edwards and Richardson (2004), Koeller et al. (2009) as well as Suikkanen et al. (2013) identified warming as a key environmental factor explaining changes in phenology of marine communities.

Hence, phenological phases are used as indicators for detecting ecological impacts of climate change on flora and fauna such as plants, migratory birds or fishes and, consequently, on ecosystems reflecting the results of manifold combinations of environmental interactions (Ahas and Aasa 2006; Ellwood et al. 2013). Changes in the timing of phenological stages of plants such as foliation, flowering, fruit ripening, discolouration and leaf fall are recognized as globally coherent ecological fingerprints of climate change. Plant development in terms of phenological stages as exemplified is a rather sensitive and, thus, ecologically significant bio-meteorological response to environmental variation (Holopainen et al. 2013; Primack and Miller-Rushing 2012) since changes in 'phenophases' serve as both forcing and inhibiting ecological processes (Noormets 2009; Parmesan 2007) across spatial scales from individuals to landscapes (Rosenzweig et al. 2008). Accordingly, shifts in plant phenology might be of importance for the implementation of mitigation and adaptation measures in compliance with regulations such as the European Habitat Directive on the one hand (van Bodegom et al. 2013) and for crop yields and food security issues on the other hand (Brown 2012; Li et al. 2014).

According to the etymologic origin of the term phenology, i.e., the ancient Greek word 'phainestai' meaning 'to appear', which was introduced by (Morren 1849), plant phenology examines annually and periodically reappearing events (phases) in growth and development of plants (Demaree and Rutishauser 2012; Schnelle 1955; Seyfert 1960). The study of the timing of recurring biological events encompasses the causes of their timing with regard to biotic and abiotic drivers. The interrelation among phases of the same or different species is called phenology (Lieth 1974). Phenology is, thus, an integrative environmental science (Schwartz 2013) covering a comprehensive methodology (Hudson and Keatley 2010). Phenological observations corroborated on the one hand that phenological phases can exhibit remarkable inter-annual variability and large spatial differences due to individual characteristics such as genes and age as well as environmental factors such as meteorological conditions at the micro and macro-scale, soil conditions, water supply, diseases, and competition. On the other hand, the seasonal development of plants is, however, mainly influenced by air temperature, photoperiod and precipitation (Clealand et al. 2012; Cook et al. 2012; Ellwood et al. 2013; Pau et al. 2011). In particular, several studies corroborated that the beginning of phenological phases, such as blooming or foliation, is closely related to air temperature (Ahas and Aasa 2006; Chmielewski et al. 2005; Chuine et al. 2000; Cook et al. 2012; Črepinšek et al. 2012; Holopainen et al. 2013; Khanduri et al. 2008; Menzel et al. 2006; Schmidt et al. 2010; Schröder et al. 2006, 2007, 2014; Sparks et al. 2000). As higher temperatures advance the course of phenological events (Kreeb 1990), phenological data reflect biological response to this feature of climate change (Braun et al. 2003) and, thus, can be used for climate biomonitoring (Gebhardt et al. 2010). Especially, spring development in the Northern Hemisphere mid latitudes mainly depends on the temperatures in winter and spring (Englert et al. 2008). Regarding the sensitivity of spring phenology of plants to warming across temporal and spatial climate gradients in independent databases, Cook et al. (2012) found good congruence, despite significant differences in species richness and geographic distribution, and concluded that this should encourage "to move beyond basic statistical diagnosis of trends and towards explicit predictions into the future" (p. 1292).

The global rise in air temperature amounted to $0.74\,°C$ in the 100-year period 1906–2005 (IPCC 2007). In Europe, even an increase of $0.95\,°C$ was observed (EEA 2008). Furthermore since air temperature data recording in 1880, the ten warmest years were all observed in the last 15 years with 2010 being the warmest year on record globally, and 2012 being the tenth warmest (Blunden and Arndt 2013)[1]. For the whole territory of Germany, there was a mean rise in air temperature of about

---

[1] http://www.climate.gov/news-features/understanding-climate/2012-state-climate-earths-sur-face-temperature.

+ 0.9 °C between the climate reference period 1961–1990 and the period 1991–2009 (Sect. 2.2.2, Fig. 2.2). The same holds true for the federal state of Hesse in central Germany (Sect. 3.2.2, Figs. 3.2, 3.3, 3.4, 3.5, 3.6) whilst due to spatial variability in some regions of Germany a warming of up to + 3.5 °C was observed (Englert et al. 2008; Schröder et al. 2007).

According to the respective climate projection and considering different greenhouse gas emission scenarios, a further rise ranging between 1.1 and 6.4 °C till 2100 has to be expected (EEA 2008; IPCC 2007). For Germany, an air temperature increase of up to 3 to 4 °C is considered until the end of the twenty-first century compared to the reference period 1971–2000 (Schröder et al. 2010), whereas in Hesse a rise between 3.2 and 3.7 °C is projected (Schönrock et al. 2012; Schröder et al. 2014).

Spatial and temporal analysis of the statistical association between phenophases and climatic drivers has to cope with the spatial incongruence of according measurement networks: In most cases, data on plant phenology are observed at different sites being apart from meteorological measurement stations (Chuine et al. 2000; Schröder et al. 2006). Hence, according data have to be coupled by help of Geographic Information Systems (GIS) and by geostatistical measures to enable mapping of phenological trends in the past and to predict future development. Accordingly, the two case studies presented in the following introduce an appropriate approach to relate phenological, i.e., data on the onset of plant phases, and meteorological observations, i.e., data on air temperatures. The first case study (Sect. 2) deals with a data analysis on the national scale (Germany), whereas the second case study (Sect. 3) allows a closer view on phenological trends by analysing data on a regional scale by example of the federal state of Hesse in central Germany.

References

Ahas R, Aasa A (2006) The effects of climate change on the phenology of selected Estonian plant, bird and fish populations. Int J Biometeorol 51:17–26

Blunden J, Arndt DS (2013) State of the climate in 2012. Bull Am Meteor Soc 94:1–258

Braun P, Brugger R, Bruns E, Clever J, Estreguil C, Flechsig M, De Groot RS, Grutters M, Harrewijn J, Jeanneret F, Martens P, Menne B, Menzel A, Sparks T (2003) European phenology network. Nature's calender on the move. Wageningen University, Netherlands

Brown I (2012) Influence of seasonal weather and climate variability on crop yields in Scotland. Int J Biometeorol 57:605–614

Caro SP, Schaper SV, Hut RA, Ball GF, Visser ME (2013) The case of the missing mechanism: how does temperature influence seasonal timing in endotherms? PLoS Biol 11(4):1–8

Chmielewski FM, Müller A, Küchler W (2005) Possible impacts of climate change on natural vegetation in Saxony (Germany). Int J Biometeorol 50:96–104

Chuine I, Cambon G, Comtois P (2000) Scaling phenology from the local to the regional level: advances from species-specific phenological models. Glob Change Biol 6(8):943–952

Clealand EE, Allen JM, Crimmins TM, Dunne JA, Pau S, Travers SE, Zavaleta ES, Wolkovich EM (2012) Phenological tracking enables positive species responses to climate change. Ecology 93(8):1765–1771

Cook BI, Wolkovich EB, Davies TJ, Ault TR, Betancourt JL, Allen JM, Bolmgren K, Clealand EE, Crimmins TM, Kraft NJB, Lancaster LT, Mazer SJ, McCabe GJ, McGill BJ, Parmesan C, Pau S, Regetz J, Salamin N, Schwartz MD, Travers SE (2012) Sensitivity of spring phenology to warming across temporal and spatial climate gradients in two independent databases. Ecosystems 15:1283–1294

Črepinšek Z, Štampar F, Kajfež-Bogataj L, Solar A (2012) The response of *Corylus avellana* L. phenology to rising temperature in north-eastern Slovenia. Int J Biometeorol 56:681–694

Demaree GR, Rutishauser T (2012) From "periodical observations" to "anthochronology" and "phenology"—the scientific debate between Adolphe Quetelet and Charles Morren on the origin of the word "phenology". Int J Biometeorol 55:753–761

Edwards M, Richardson AJ (2004) Impact of climate change on marine pelagic phenology and trophic mismatch. Nature 430: 881–884

EEA (European Environment Agency) (2008) Impacts of Europe's changing climate—2008 indicator-based assessment. EEA Report 4/2008, Copenhagen

Ellwood ER, Temple SA, Primack RB, Bradley NL, Davis CC (2013) Record-breaking early flowering in the Eastern United States. PLoS ONE 8(1):1–9

Englert C, Pesch R, Schmidt G, Schröder W (2008) Analysis of spatially and seasonally varying plant phenology in Germany. In: Car A, Griesebner G, Strobl J (eds) Geospatial Crossroads @ GI_Forum '08: proceedings of the geoinformatics forum Salzburg. Wichmann, Heidelberg, pp 81–89

Foden WB, Butchart SHM, Stuart SN, Vié J-C, Akçakaya HR, et al (2013) Identifying the world's most climate change vulnerable species: a systematic trait-based assessment of all birds, amphibians and corals. PLoS ONE 8(6):1–13

Gebhardt H, Rammert U, Schröder W, Wolf H (2010) Klima-Biomonitoring: Nachweis des Klimawandels und dessen Folgen für die belebte Umwelt. Umweltwiss Schadst Forsch 22(1):7–19

Holopainen J, Helama S, Lappalainen H, Gregow H (2013) Plant phenological records in northern Finland since the eighteenth century as retrieved from databases, archives and diaries for biometeorological research. Int J Biometeorol 57:423–435

Hudson IL, Keatley MR (eds) (2010) Phenological research: methods for environmental and climate change analysis. Springer, New York

IPCC (International Panel on Climate Change) (2007) Climate change 2007. Synthesis report. Geneva, 52 p

Khanduri VP, Sharma CM, Singh SP (2008) The effects of climate change on plant phenology. Environmentalist 28:143–147

Koeller P, Fuentes-Yaco C, Platt T, Sathyendranath S, Richards A, Ouellet P, Orr D, Skúladóttir U, Wieland K, Savard L, Aschan M (2009) Basin-scale coherence in phenology of shrimps and phytoplankton in the North Atlantic ocean. Science 324(5928):791–793

Kreeb KH (1990) Methoden zur Pflanzenökologie und Bioindikation. Fischer, Stuttgart

Li Z, Yang P, Tang H, Wu W, Yin H, Liu Z, Zhang L (2014) Response of maize phenology to climate warming in Northeast China between 1990 and 2012. Reg Environ Change 14:39–48

Lieth H (ed) (1974) Phenology and seasonality modelling. Springer, New York

Menzel A, Sparks TH, Estrella N, Koch E et al (2006) European phenological response to climate change matches the warming pattern. Glob Change Biol 12(10):1969–1976

Morren C (1849) Le globe, le temps et la vie. Bulletins de l'Académie royale des Sciences, des Lettres et des Beaux-Arts de Belgique XVI(2):660–684

Mouillot D, Bellwood DR, Baraloto C, Chave J, Galzin R et al (2013) Rare species support vulnerable functions in high-diversity ecosystems. PLoS Biol 11(5):1–11

Noormets A (ed) (2009) Phenology of ecosystem processes: applications in global change research. Springer, New York

Parmesan C (2007) Influences of species, latitudes and methodologies on estimates of phenological response to global warming. Glob Change Biol 13:1860–1872

Pau S, Wolkovich EM, Cook BI, Davies JT, Kraft NJB, Blomgren K, Betancourt JL, Clealand EE (2011) Predicting phenology by integrating ecology, evolution and climate science. Glob Change Biol 17:3633–3643

Polgar CA, Primack RB, Williams E, Stichter S, Hitchcock C (2013) The effect of temperature on the timing of the adult flight period of Lycaenid butterflies in Massachusetts. Biol Conserv 160:25–31

Primack RB, Miller-Rushing AJ (2012) Uncovering, collecting and analyzing records to investigate the ecological impacts of climate change: a template from Thoreau's Concord. BioScience 62:170–180

Rosenzweig C, Karoly D, Vicarelli M, Neofotis P, Wu Q, Casassa G, Menzel A, Root TL, Estrella N, Seguin B, Tryjanowski P, Liu C, Rawlins S, Imeson A (2008) Attributing physical and biological impacts to anthropogenic climate change. Nature 453:353–357

Schmidt G, Holy M, Pesch R, Schröder W (2010) Changing plant phenology in Germany due to the effects of global warming. Int J Climate Change 2(2):73–84

Schnelle F (1955) Pflanzen-Phänologie. Geest and Portig, Leipzig

Schönrock S, Schmidt G, Schröder W (2012) Climate Biomonitoring: Impacts of climate change on plant phenology in Hesse (Germany). In: Olabi AG, Benyounis KY (eds) Environment and clean technologies. Proceedings of the 5th international conference on sustainable energy and environmental protection, SEEP 2012, 05th–08th June 2012, DCU, Dublin, Rep. Ireland, pp 44–49

Schröder W, Pesch R (2011) Mapping carbon sequestration in forests at the regional scale. A climate biomonitoring approach by example of Germany. Environ Sci Europe 23(31):1–16

Schröder W, Schmidt G (2014) Modelling potential malaria spread in Germany by use of climate change projections. A risk assessment approach coupling epidemiologic and geostatistical measures. (SpringerBriefs). Springer, Heidelberg

Schröder W, Schmidt G, Hasenclever J (2006) Geostatistical analysis of data on air temperature and plant phenology from Baden-Württemberg (Germany) as a basis for regional scaled models of climate change. Environ Monit Assess 130(1–3):27–43

Schröder W, Englert C, Schmidt G (2007) Phänologische Änderungen bei Obstbäumen und anderen Pflanzen sowie weitere mögliche Folgen des Klimawandels für die Landwirtschaft. LandInfo 5:1–15

Schröder W, Pesch R, Schmidt G (2010) Klimawandel. In: Schröder W, Fränzle O, Müller F (eds) Handbuch der Umweltwissenschaften. Grundlagen und Anwendungen der Ökosystemforschung, (18. Erg. Lfg., Kap. VI-1.4). Wiley VCH, Weinheim, pp 1–23

Schröder W, Schmidt G, Schönrock S (2014) Modelling and mapping of plant phenological stages as bio-meteorological indicators for climate change. Environ Sci Europe 24(5):1–13

Schwartz MD (2013) Phenology: an integrative environmental science, 2nd edn. Springer, New York

Seyfert F (1960) Phänologie. Ziemsen, Wittenberg

Sparks TH, Jeffree EP, Jeffree CE (2000) An examination of the relationship between flowering times and temperature at the national scale using long-term phenological records from the UK. Int J Biometeorol 44:82–87

Suikkanen S, Pulina S, Engström-Öst J, Lehtiniemi M, Lehtinen S, Brutemark A (2013) Climate change and eutrophication induced shifts in northern summer plankton communities. PLoS ONE 8(6):1–10

van Bodegom PM, Verboom J, Witte JPM, Vos CC, Bartholomeus RP, Geertsema W, Cormont A, van der Veen M, Aerts R (2013) Synthesis of ecosystem vulnerability to climate change in the Netherlands shows the need to consider environmental fluctuations in adaptation measures. Reg Environ Change 14 (3):933–942

Chapter 2
Case Study 1: Phenological Trends in Germany

Abstract The rise of air temperature as one result of climate change affects the developmental stages of plants. In Germany, systematic observation of plant phenology development is established by help of International Phenological Gardens (IPGs) and by the phenological network of the German Meteorological Service (DWD) with up to 6500 observation sites for 270 phenological plant phases. *Case study 1* aims at quantifying the statistical association between distinct plant phenological phases ('phenophases') and measured air temperatures in Germany. This should help in future projections on possible impacts of climate change on plant development and distribution. Accordingly, data on mean annual air temperatures and country-wide observations on 6 phases indicating different plant phenological seasons were analysed by means of regression analysis. Within a Geographic Information System (GIS), Regression Kriging was applied for mapping the development of plant phenology in the past and also in future by using projected temperature data for the climate reference periods 1991–2020, 2021–2050, and 2051–2080 derived from two different climate models (REMO, WettReg) and two emission scenarios (A1B, B1). The results showed already for the comparison of the past climate periods 1961–1990 and 1991–2005 a distinct shift of phenological onset towards the beginning of the year by about 9 days in average for all 6 phases investigated. The strongest shift was observed for hazel bloom advancing 13 days to the beginning of the year. In future, a shift of up to 33 days was calculated comparing data of 1961–1990 and 2051–2080. Since WettReg projections assume a moderate temperature rise, the projected phenological shifts were not that pronounced compared to REMO scenarios. Hence for WettReg B1, a shift by only 17 days was calculated for the beginning of hazel bloom. The strongest relationship between annual air temperatures and phase onset was found for phenophases in spring and, accordingly, the shifts in the beginning of phenophases indicating the spring season were most intense. The algorithms describing the statistical relation between temperature rise and phenology development were integrated into the "Technical Information System on Climate Change and Adaptation Strategies" (Fachinformationssystem Klimawandel und Anpassung, FISKA) which was initiated by the Federal Environment Agency and administered by the Competence Centre on Climate Impacts and Adaptation (KomPass). For implementation, so called 'calculation engines' were developed for all 6 indicator phases, each calculating the beginning of the respective phenological phase in days after New Year based on the respective mean annual air temperatures. That allows for future assessments when improved

G. Schmidt et al., *Plant Phenology as a Biomonitor for Climate Change in Germany*, SpringerBriefs in Environmental Science, DOI 10.1007/978-3-319-09090-0_2

emission scenarios or climate models are available in order to develop well adapted mitigation measures considering environmental, agricultural and economic issues emerging from changes in plant growth and distribution.

Keywords German adaptation action plan · Phenological monitoring · Climate change impacts · Mitigation measures · Expert knowledge system

2.1 Background and Goals

According to the recently published Summary for Policymakers (SPM)[1], condensing the most important findings of the latest IPCC report in 2013, it can be stated that "each of the last three decades has been successively warmer at the Earth's surface than any preceding decade since 1850. In the Northern Hemisphere, 1983–2012 was likely the warmest 30-year period of the last 1400 years" (IPCC 2013, p. 3). Even in 2005, the German government claimed to limit the global temperature increase to $+2\,°C$ compared to the pre-industrial state[2]. This statement, in fact, relies on recommendations of the German Advisory Council on Global Change (Wissenschaftlicher Beirat der Bundesregierung Globale Umweltveränderungen (WBGU) released already in 1995[3] (Jaeger and Jaeger 2011).

However, even a rise of $+2\,°C$ will induce various impacts on environment, economy, and society. Hence, the German adaptation strategy on climate change (Deutsche Anpassungsstrategie—DAS)[4] describes potential impacts, goals, possible conflicts, and adaptation measures including climate risk assessment and management as well as disaster risk reduction strategies for national policies and programmes. For putting this into action, 13 fields of investigation were defined comprising environmental impacts as well as socio-economic issues. However, in contrast to a report published by the European Environmental Agency (EEA 2008), the DAS provides no spatial differentiation of climate change induced observed or projected environmental impacts in Germany. Thus, the implementation of the "Expert Information System on Climate Change and Adaption" (FISKA) was initiated by the German Federal Environment Agency and administered by the Competence Centre on Climate Impacts and Adaptation (KomPass)[5]. FISKA shall provide the governmental institutions with basic information and models on climate change impacts for the development and accomplishment of adaption and mitigation strategies. A main objective was to implement an information system providing certain impact models that make it possible to provide spatially differentiated data and models to enable preventive assessment of possible impacts of climate change (Schröder et al. 2010). Accordingly, the approach based on a Geographic

---

[1] http://www.de-ipcc.de/_media/WGIAR5_SPM_brochure_en.pdf.

[2] http://www.bmu.de/N35742.

[3] http://www.wbgu.de/en/special-reports/sr-1995-co2-reduction.

[4] http://www.bmu.de/N42783.

[5] http://www.umweltbundesamt.de/en/topics/climate-energy/climate-change-adaptation/kompass.

Information System (GIS). The GIS prototype was provided with exemplary data and calculation kernels that enable modelling certain impacts of climate change for different spatial extents and periods (Schmidt et al. 2010). One of these kernels allows projection of plant phenology development (Sect. 2.1) by example of six different phenophases (Sect. 2.2.1). For mapping spatial patterns of phenology development (Sect. 2.4), Regression Kriging (Sect. 2.3) was applied within a GIS environment relying on the statistical association between air temperatures and phenophases' onset. Accordingly, the derived regression functions were applied for different temperature grids based on two climate models (REMO, WettReg) regarding two IPCC emission scenarios (B1, A1B) and covering climate periods in the past (1961–1990, 1991–2005) and in the future (1991–2020, 2021–2050, 2051–2080) (Sect. 2.2.2).

2.2   Materials

For mapping the developmental stages of plants in Germany certain phenophases were correlated with air temperature measurements for two observation periods in the past (1961–1990, 1991–2005). The resulting regression equations were also used for depicting possible future development of phenophase onset regarding the respective climate projections.

2.2.1   Phenology Data

In Germany, systematic observation of plant phenology is established by international phenological gardens[6] and by the phenological network of the German Meteorological Service (DWD)[7] with up to 6500 observation sites (Fig. 2.1, right). For *case study 1*, phenological observations on 6 phases indicating different plant phenological seasons were analysed by means of regression analysis (Sect. 2.3): hazel (beginning of flowering pre-spring), forsythia (beginning of flowering, first spring), apple (beginning of flowering, full spring), elder (first bloom, early summer), large-leaved lime (first bloom, midsummer), and apple (fruit maturity, late summer) (Table 2.1). Data on phenology observations were available from 1961 to 2005.

The phenological observations at each site were conducted two or three times a week within a defined area with a radius of 5 km by volunteers according to a guideline published by the German Weather Service (DWD 1991). Phenological data were provided as vector data sets (point layer) and were processed within a GIS environment. For analysis, all those observation sites were considered where for at

---

[6] http://www.agrar.hu-berlin.de/fakultaet/departments/dntw/agrarmet/phaenologie/ipg.

[7] http://www.dwd.de/bvbw/appmanager/bvbw/dwdwwwDesktop?_nfpb=true&_pageLabel=_dwdwww_klima_umwelt_phaenologie&activePage=&_nfls=false.

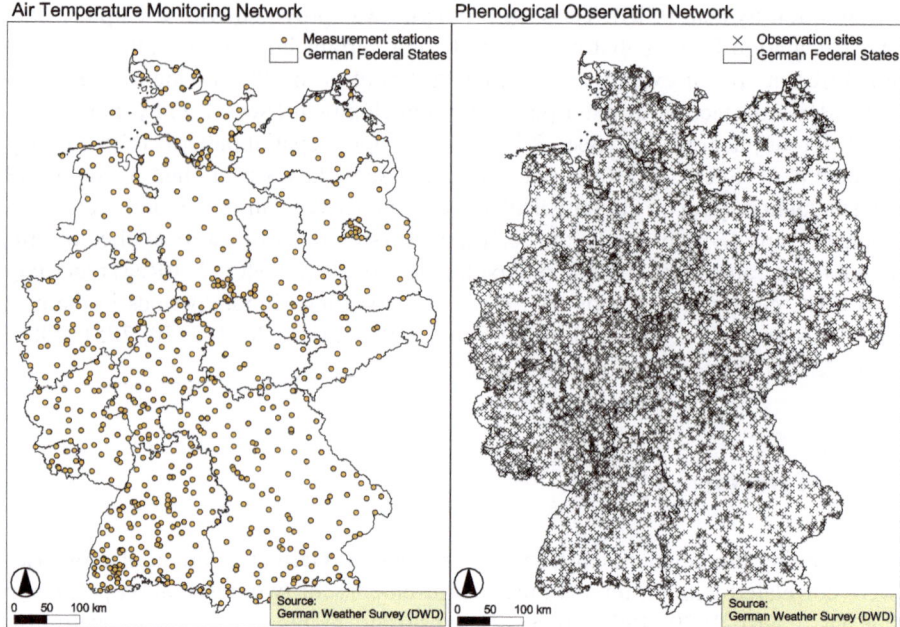

Fig. 2.1 Measurement networks on air temperature (*left*) and plant phenology (*right*) in Germany, both maintained by the German Weather Service (DWD)

Table 2.1 Investigated plant phases in Germany

Code	Plant	Common name	Phase	Phenological season
1	*Corylus avellana*	Hazel	Beginning of flowering	Pre-spring
6	*Forsythia suspensa*	Forsythia	Beginning of flowering	First spring
62	*Malus domestica*	Apple	Beginning of flowering	Full spring
18	*Sambucus nigra*	Black elder	Beginning of flowering	Early summer
64	*Tilia platyphyllos*	Large-leaved lime	Beginning of flowering	Midsummer
109	*Malus domestica*	Apple, early ripening	Fruit ripe for picking	Late summer

least 80 % of the years within an observation period records were available. Accordingly, for the climate reference period 1961–1990 there had to be at least 24 observations at a particular site and for the period 1991–2005 at least 12 observations. By performing T-tests and by comparing descriptive statistical measures it could be assured that there were no significant differences between both data cohorts comparing sites with a complete temporal coverage and sites with an 80 % coverage.

2.2.2   Data on Air Temperatures

Data on mean monthly air temperatures were provided by the German Meteorological Service (DWD) for the climate reference period 1961–1990 and the period

1991–2009. The data are based on about 670 measurement stations spread across Germany (Fig. 2.1, left). These local measurements were transformed to surface maps (Fig. 2.2) by means of Regression Kriging (Sect. 2.3). This was performed by correlating temperature values and altitudes of the respective measurement sites. The correlation coefficients indicated high ($r = 0.75$–0.83) and significant relationships. In the GIS, the regression model was used to calculate high resolution long-term annual temperature maps for both periods based on the global digital elevation model GLOBE[8] (spatial resolution $= 1 \times 1$ km^2).

Temperature data for future climate reference periods 1991–2020, 2021–2050, and 2051–2080 were derived from the climate projections REMO (Regional Model; Max Planck Institute for Meteorology) (Jacob et al. 2008) and WettReg (Weather Condition-based Regionalisation Method; Climate & Environment Consulting Potsdam) (Spekat et al. 2006) which are based on the global ECHAM5 climate model. ECHAM5 is the 5th generation of the ECHAM general circulation model. Depending on the configuration, the model resolves the atmosphere up to 10 hPa for tropospheric studies or up to 0.01 hPa for middle atmosphere studies (often referred to as MAECHAM5) (Roeckner et al. 2006). Both REMO and WettReg projections were applied considering two emission scenarios: Scenario A1B assumes a rapid economic growth, a global population reaching a number of 9 billion in 2050 and then showing a gradual decline, the quick spread of new and efficient technologies and a balanced use of all energy sources. Scenario B1 assumes a more integrated and more ecologically healthy world with rapid economic growth as in A1. Further assumptions include, rapid changes towards a service and information economy, a population increasing up to 9 billion in 2050 and then declining as in scenario A1B, although with reductions in material intensity and the introduction of clean and resource efficient technologies as well as an emphasis on global solutions to economic, social and environmental stability (IPCC 2007). The spatial resolution of both REMO and WettReg grids was 12×12 km^2.

Compared with WettReg (Fig. 2.3), REMO (Fig. 2.4) estimates higher temperatures and stronger increase. Regarding climate model WettReg and emission scenario B1, which supposes a more moderate rise of greenhouse gases, the overall air temperature mean should increase by 1.0 °C (Fig. 2.3, lower row) comparing the periods 1991–2020 and 2051–2080 whereas for the scenario A1B air temperatures are expected to increase by about 1.7 °C (Fig. 2.3, upper row).

Regarding climate model REMO and emission scenario B1, the overall air temperature mean should increase by 1.2 °C (Fig. 2.4, lower row) comparing the periods 1991–2020 and 2051–2080 whereas for the scenario A1B the modelled temperature increase was 2.2 °C (Fig. 2.4, upper row).

---

[8] http://www.ngdc.noaa.gov/mgg/topo/globe.html.

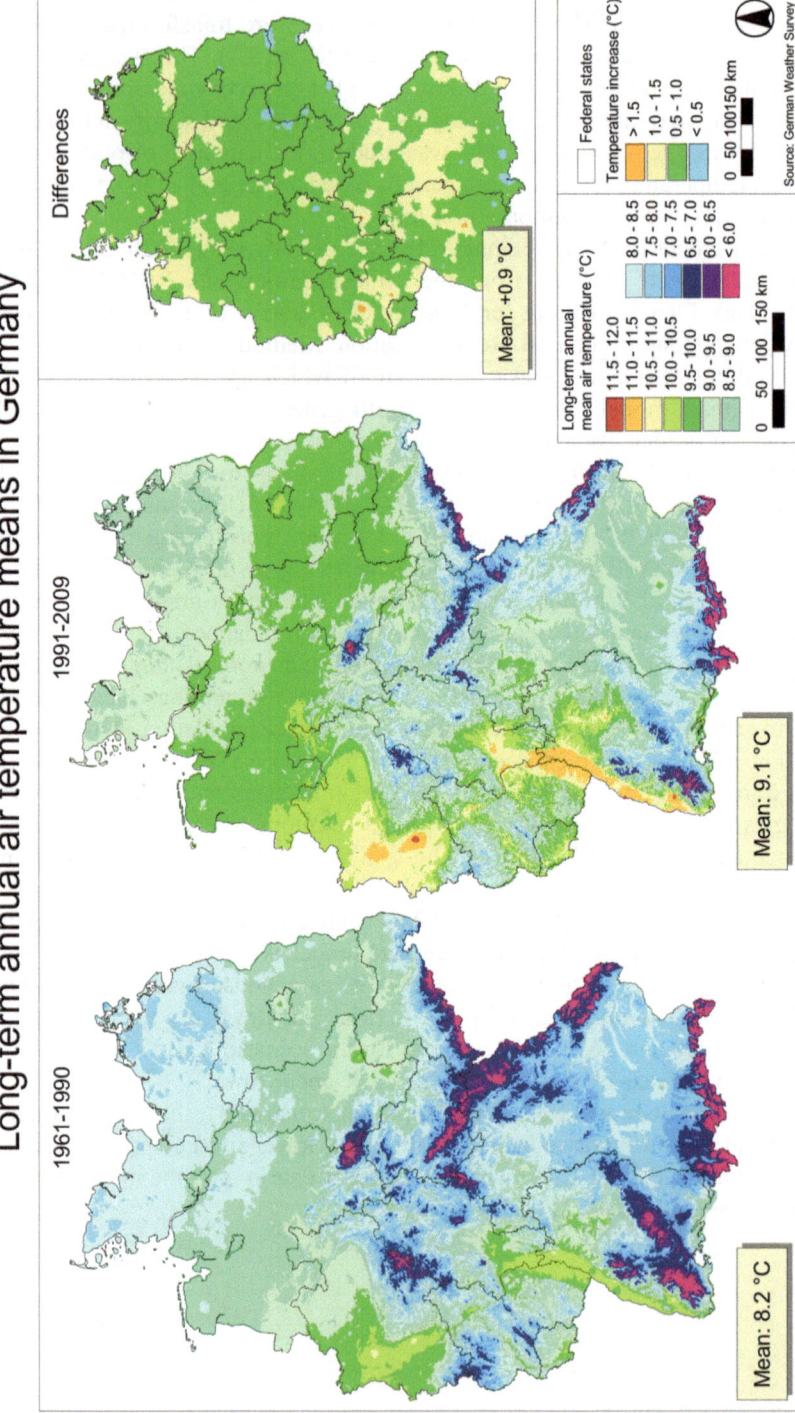

Fig. 2.2 Long-term annual means on air temperatures in Germany for the climate reference period 1961–1990 (*left*) and the period 1991–2009 (*centre*) as well as according differences between both periods (*right*)

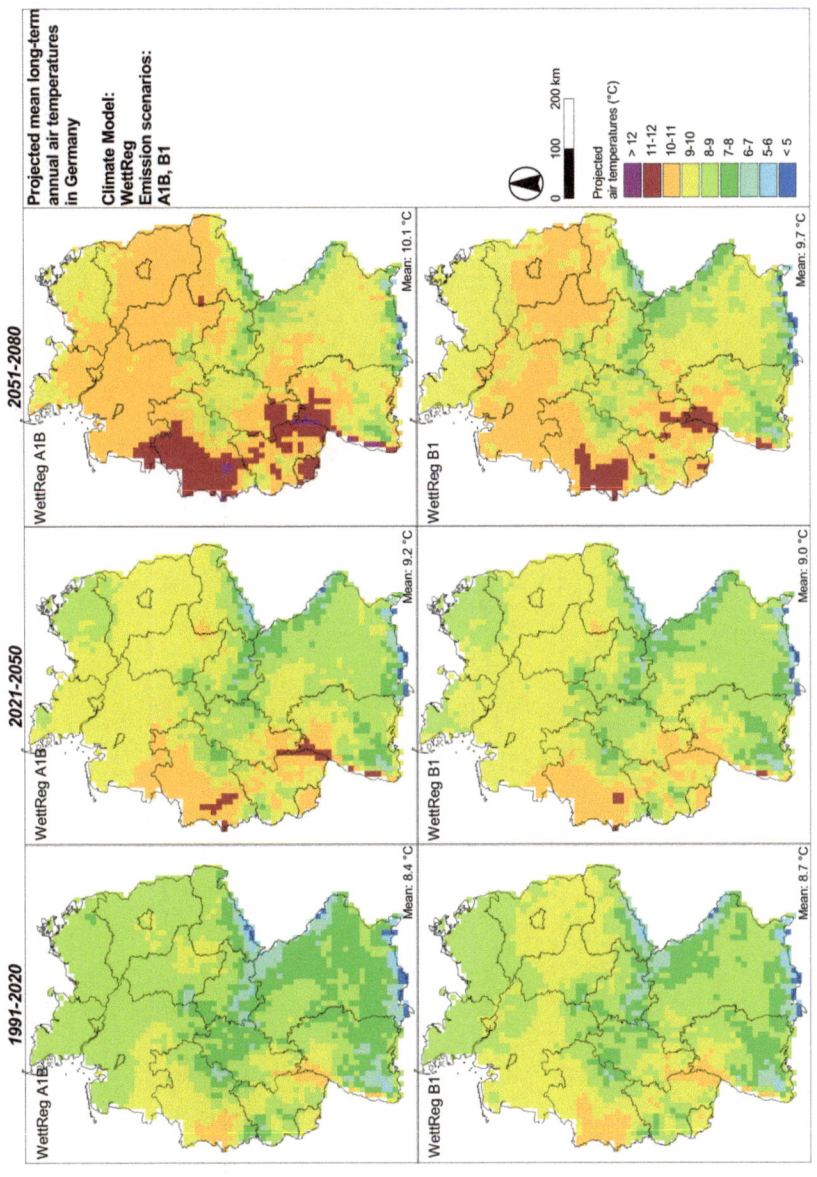

Fig. 2.3 Projected long-term annual means on air temperatures in Germany for the climate periods 1991–2020 (*left*), 2021–2050 (*centre*) and period 2051–2080 (*right*) according to climate model WettReg and considering emission scenario A1B (*upper row*) and B1 (*lower row*)

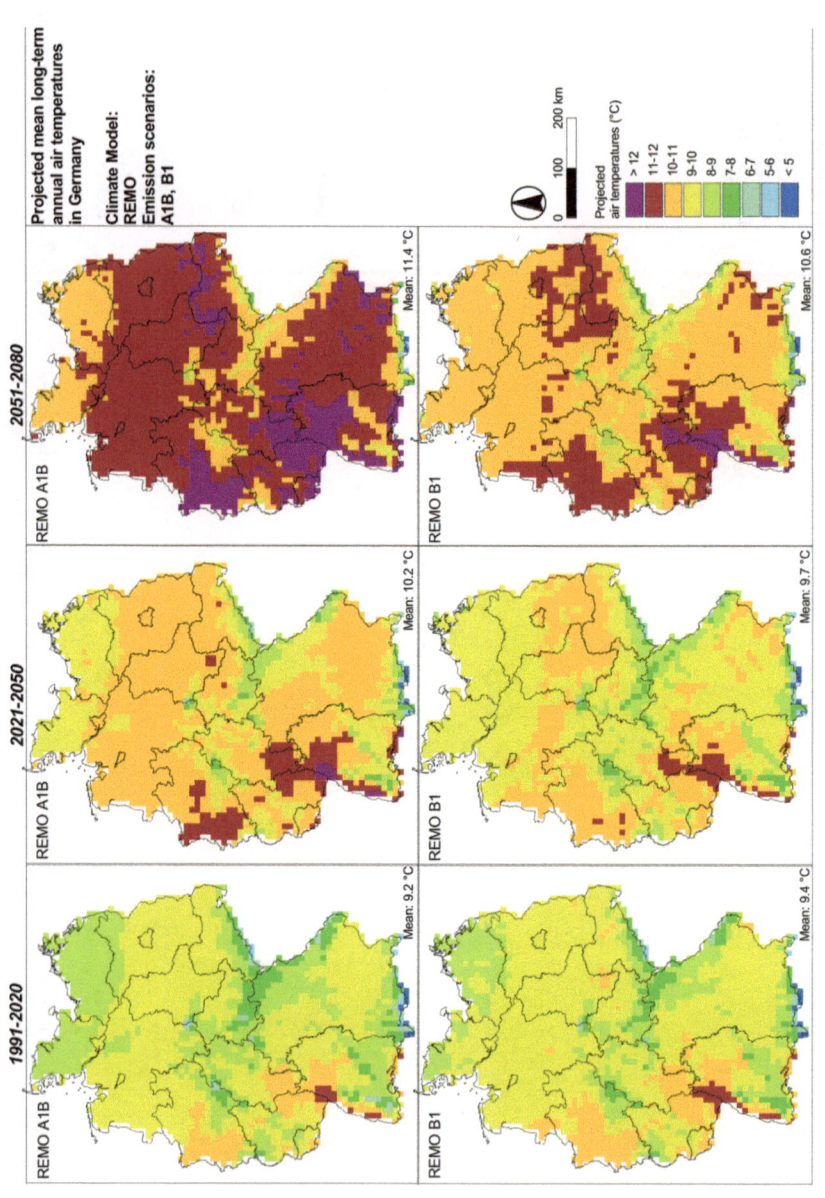

Fig. 2.4 Projected long-term annual means on air temperatures in Germany for the climate periods 1991–2020 (*left*), 2021–2050 (*centre*) and period 2051–2080 (*right*) according to climate model REMO and considering emission scenario A1B (*upper row*) and B1 (*lower row*)

Table 2.2 Data used for ecological land classification. (According to Schröder and Schmidt 2001)

Map title	Period/state	GIS layer	Source
Potential natural vegetation	1998	1	BfN
Soil texture	2000	1	BGR
Altitude above sea level	1996	1	UNEP
Mean of monthly global radiation March–November	1981–1999	9	DWD
Mean of monthly evaporation January–December	1961–1990	12	DWD
Mean of monthly precipitation January–December	1961–1990	12	DWD
Mean of monthly air temperature January–December	1961–1990	12	DWD

BfN Federal Agency for Nature Conservation, *BGR* Federal Institute for Geosciences and Natural Resources, *GIS* Geographical Information System, *UNEP* United Nations Environmental Programme, *DWD* German National Meteorological Service

2.2.3   Ecological Land Classification

For a spatially differentiated view on the respective phenological shifts in Germany the mapping results derived by Regression Kriging (Sect. 2.3) were intersected in a GIS with a map on Germany's ecoregions (Schmidt 2002; Schröder and Schmidt 2001). Ecological land classifications were computed based on data which represent several interacting factors which may be of importance for natural processes such as the developmental stages of plants. The ecological land classification used for *case study 1* was calculated from the data listed in Table 2.2 applying Classification and Regression Trees (CART) (Breiman et al. 1984). The respective computations yielded a map (Fig. 2.5) illustrating the patterns of spatially discriminated and ecologically defined land classes which synonymously are called 'natural land classes', 'ecoregions' or 'landscapes' (Schröder 2006, Schröder et al. 2006b). All ecoregions are itemised with regard to the values or the statistical distribution of values of 48 ecological characteristics which were used to generate the land classes by use of CART (Schröder and Schmidt 2001).

2.3   Methods

For generating maps depicting the mean beginning of certain phenological phases in Germany, Pearson's correlation coefficients between long-term annual air temperature and each phenological phase were calculated for both periods 1961–1990 and 1991–2005. For each observation site of the phenological network the according temperature value was extracted from the temperature grid (Figs. 2.2, 2.3, 2.4) averaged for each period. Like for processing the air temperature maps for the climate periods in the past (Sect. 2.2.2), Regression Kriging (Hengl et al. 2007; Odeh et al. 1995) was performed to derive maps on the phenological development: Regression functions for both periods in the past (1961–1990, 1991–2005) were applied in the GIS to map the beginning of the respective phenological phase. Residual maps on the differences between measured and modelled temperatures were calculated by

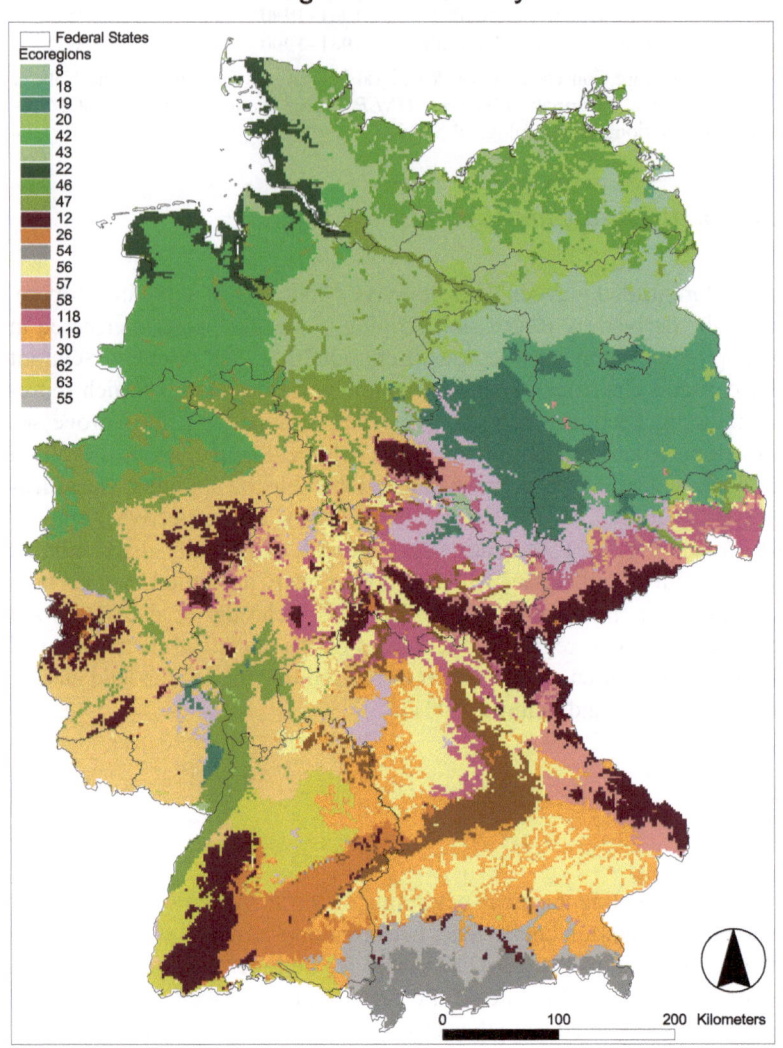

Fig. 2.5 Ecoregions of Germany calculated by CART (Classification and Regression Trees) from the data in Table 2.1. (According to Schröder and Schmidt 2001)

Table 2.3 Correlation coefficients (Pearson) between long-term mean annual air temperatures and phase onset for both observation periods (1961–1990, 1991–2005)

Code	Common name	Phase	r (1961–1990)	r (1991–2005)
1	Hazel	Flowering	−0.80	−0.54
6	Forsythia	Flowering	−0.85	−0.81
18	Black elder	Flowering	−0.70	−0.69
62	Apple	Flowering	−0.76	−0.78
64	Large-leaved lime	Flowering	−0.66	−0.56
109	Apple	Fruit ripening	−0.66	−0.47

means of Ordinary Kriging based on auto-correlation functions determined by use of variogram analysis and subtracted from the Regression Kriging maps to account for over- or underestimations (Hengl et al. 2007; Romić et al. 2007). The quality of the surface estimations was assessed by means of cross-validation (Johnston et al. 2001). The resulting maps on the beginning of the respective phenological indicator phase had a spatial resolution of 5×5 km^2 (Sect. 2.4.1). Future development of phenology was modelled by applying the derived regression function for the period 1991–2005 for each phenological phase to the projected temperature grids provided by REMO and WettReg climate models (Sect. 2.2.2). According to the spatial resolution of the projected temperature grids, each map had a resolution of 12×12 km^2. Finally, each map on the respective phenological indicator phase in each period was intersected with a map on German ecoregions (Sect. 2.2.3) to identify regional differences in the beginning of the plant phases (Sect. 2.4.2).

2.4 Results

Considering both climate models (WettReg, REMO) and both emission scenarios (A1B, B1) described by the IPCC (2007), for each of the six phenological phases (Sect. 2.2.1) two maps for the past development (1961–1990, 1991–2005) and 12 maps for the future development (1991–2020, 2021–2050 and 2051–2080) were calculated by means of Regression Kriging (Sect. 2.3).

2.4.1 Phenological Development in the Past and Future

For the two observation periods in the past (1961–1990, 1991–2005) Pearson's correlation coefficients were calculated revealing medium to high statistical correlation between mean air temperatures and phase onset, the values varied between −0.47 (phase 109, 1991–2005) and −0.85 (phase 6, 1961–1990) (Table 2.3).

All correlations were statistically significant ($p < 0.001$). Exemplarily, Fig. 2.6 depicts the results of the regression analysis for phase 64 (flowering of large-leaved lime) for the period 1991–2005.

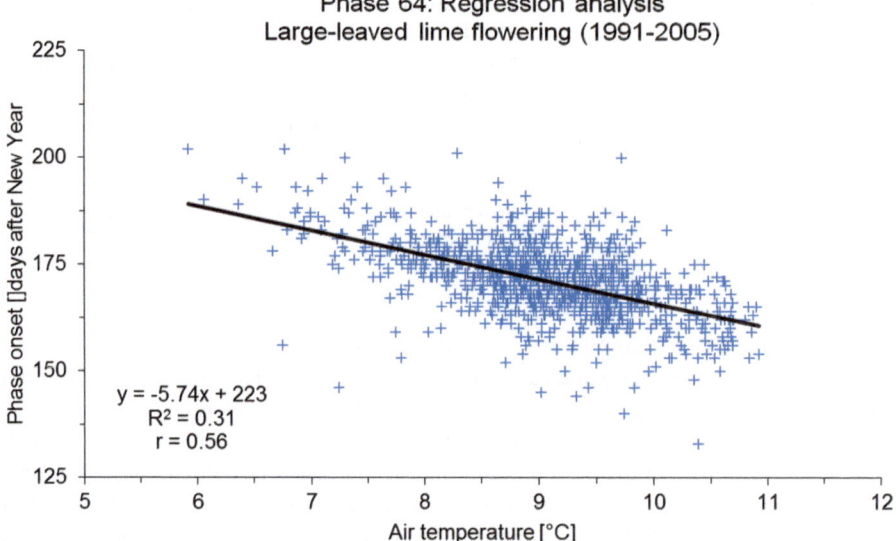

Fig. 2.6 Regression analysis for the statistical association between the beginning of flowering of *Tilia platyphyllos* (large-leaved lime) and air temperatures in Germany for the climate period 1991–2005

As described in Sect. 2.3, the respective regression functions were used within a GIS environment to calculate grids for both observation periods in the past (1961–1990, 1991–2005) and for both climate models (WettReg, REMO) and emission scenarios (B1, A1B) for the climate reference periods 1991–2020, 2021–2050, and 2051–2080. Exemplarily, Figs. 2.7, 2.8, 2.9, 2.10 depict the results of Regression Kriging for phase 64. In each of these figures, the two upper maps depict the phase onset for both observation periods (1961–1990, 1991–2005) showing the calculated grids and the observation sites where the blooming of *Tilia platyphyllos* was recorded in at least 80 % of the years of the respective period (Sect. 2.2.1). Additionally, the lower three maps show the grids calculated by applying the regressions equation derived for the period 1991–2005 to the respective air temperature grids projected by the respective climate model (WettReg, REMO) and emission scenario (B1, A1B) (Sect. 2.2.2).

Comparing the maps for each climate projection one can see that *early* phase onset coincides with high air temperatures in Germany (Fig. 2.2) occurring especially in the Upper Rhine valley in the southwest of Germany and in the Westphalian lowlands in the northwest. Accordingly, *late* flowering of large-leaved lime was observed in the higher and lower mountain ranges around the Alps, the Black Forest and the Bavarian Forest, for example. On average, in the climate reference period 1961–1990 the flowering of large-leaved lime (*Tilia platyphyllos*) began 178 days after New Year, whereas for 1991–2005 there was a shift of 1 week to the beginning of the year (171 days). For the first projection period (1991–2020) quite the same spatial distribution of phase onset compared to the recent observation period (1991–

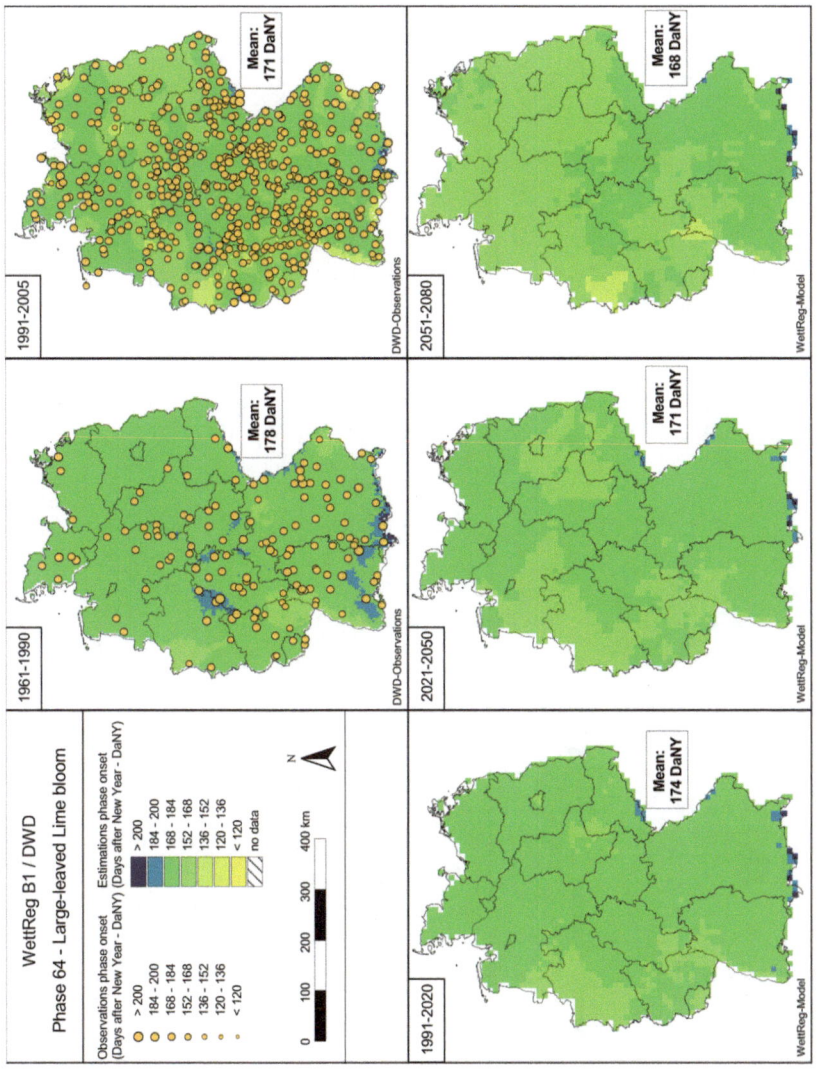

Fig. 2.7 Observed (1961–1990, 1991–2005) and projected (1991–2020, 2021–2050, 2051–2080) onset of flowering of *Tilia platyphyllos* (large-leaved lime) in Germany according to climate model WettReg and emission scenario B1

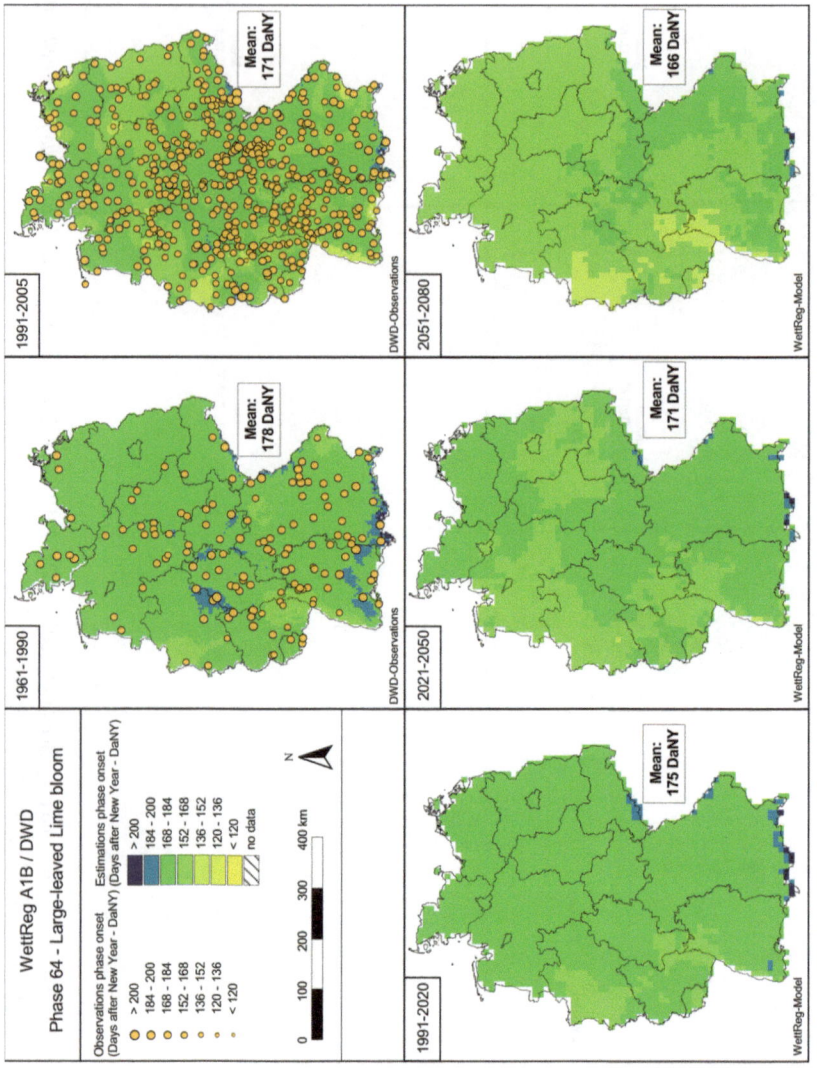

Fig. 2.8 Observed (1961–1990, 1991–2005) and projected (1991–2020, 2021–2050, 2051–2080) onset of flowering of *Tilia platyphyllos* (large-leaved lime) in Germany according to climate model WettReg and emission scenario A1B

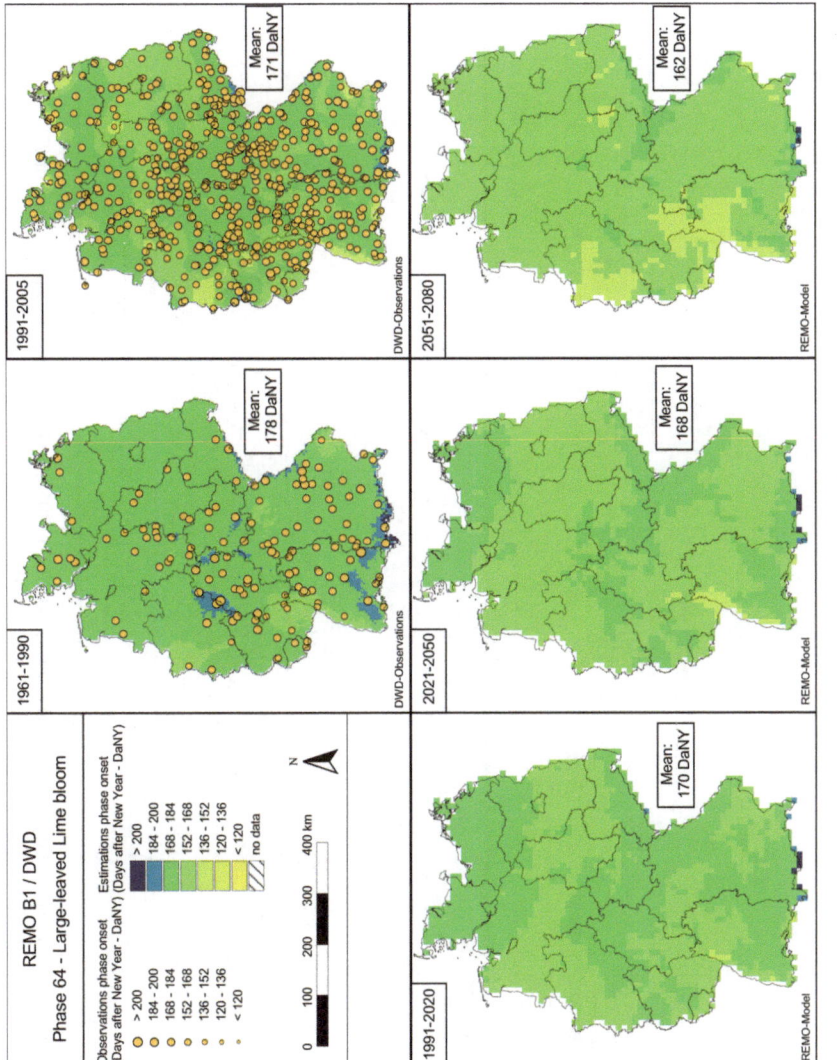

Fig. 2.9 Observed (1961–1990, 1991–2005) and projected (1991–2020, 2021–2050, 2051–2080) onset of flowering of *Tilia platyphyllos* (large-leaved lime) in Germany according to climate model REMO and emission scenario B1

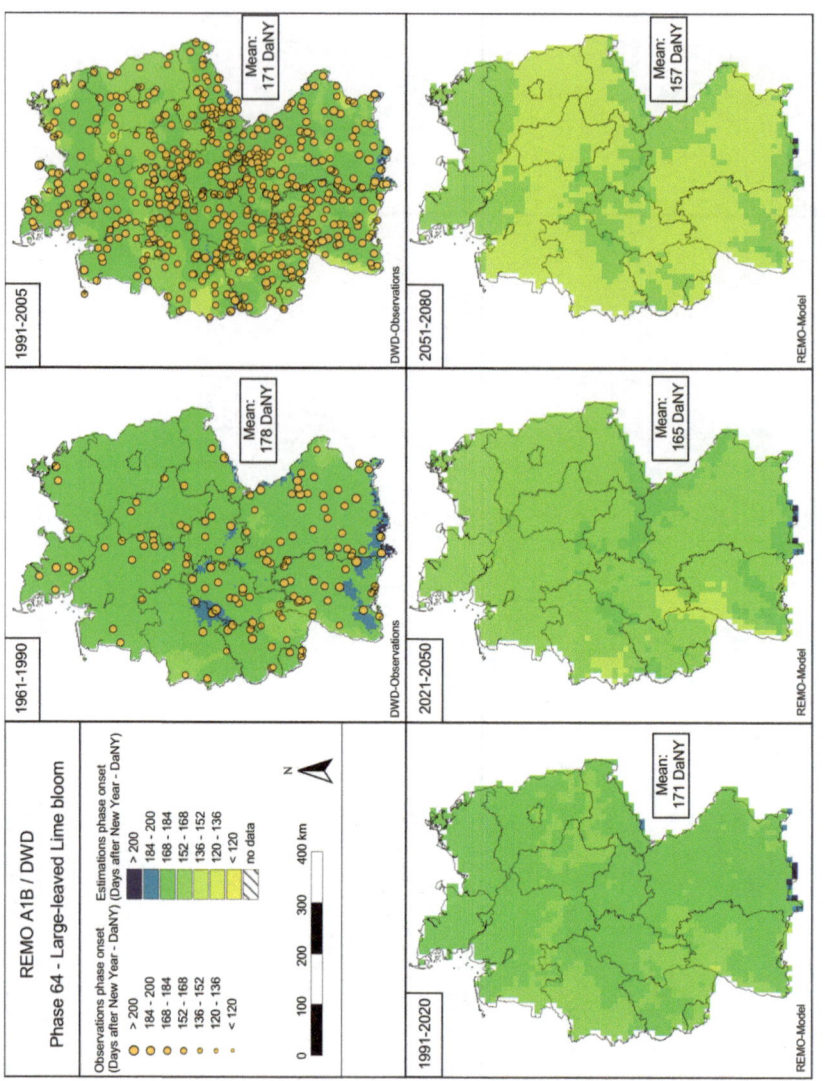

Fig. 2.10 Observed (1961–1990, 1991–2005) and projected (1991–2020, 2021–2050, 2051–2080) onset of flowering of *Tilia platyphyllos* (large-leaved lime) in Germany according to climate model REMO and emission scenario A1B

Table 2.4 Mean phase onset (days after New Year) for all analysed periods and climate scenarios and resulting differences (days) between the climate reference periods 1961–1990 and 2051–2080

Scenario, period	Phase					
	1	6	62	18	64	109
DWD, 1961–1990	61	96[a]	128	157[a]	178	220
DWD, 1991–2005	48	85	119	149	171	212
REMO A1B, 1991–2020	48	85	118	148	171	211
REMO A1B, 2021–2050	39	75	111	142	165	205
REMO A1B, 2051–2080	28	64	103	133	157	198
REMO B1, 1991–2020	46	83	117	147	170	210
REMO B1 2021–2050	43	79	114	145	168	208
REMO B1 2051–2080	35	71	108	139	162	203
WettReg A1B, 1991–2020	54	92	123	153	175	216
WettReg A1B, 2021–2050	48	85	118	148	171	211
WettReg A1B, 2051–2080	40	76	112	142	166	206
WettReg B1, 1991–2020	52	89	121	151	174	214
WettReg B1, 2021–2050	49	86	119	149	171	212
WettReg B1, 2051–2080	44	80	115	145	168	208
REMO A1B, differences	−33	−32	−25	−24	−21	−22
REMO B1, differences	−26	−25	−20	−18	−16	−17
WettReg A1B, differences	−21	−20	−16	−15	−12	−14
WettReg B1, differences	−17	−16	−13	−12	−8	−12

[a] No data available for Eastern Germany

2005) was calculated. WettReg scenarios even projected a later onset (174 and 175 days) of lime bloom compared to 1991–2005. Obviously, measured air temperatures in the last decades already exceeded the anticipations made for the WettReg scenarios. Finally, for the climate period 2051–2080 the phase onset should shift in worst case (REMO A1B) another 2 weeks to the beginning of the year (157 days after New Year) compared to the period 1991–2005 and 3 weeks compared to the climate reference period 1961–1990. In comparison, it can be stated that due to higher air temperatures phase onset for *Tilia platyphyllos* was earlier in A1B scenarios compared to the B1 scenarios. Additionally, projections relying on REMO scenarios showed an earlier beginning of flowering than those relying on WettReg scenarios as the latter predicts a more moderate air temperature increase (Sect. 2.2.2).

Table 2.4 summarizes averaged phase onsets for all observed (1961–1990, 1991–2005) and projected climate periods (1991–2020, 2021–2050, 2051–2080) considering both climate models (WettReg, REMO) and emission scenarios (A1B, B1).

Overall, the onset of the six investigated phenological phases showed a distinct shift towards the beginning of the year varying between 7 days (phase 64) and 13 days (phase 1). Considering hazel bloom (phase 1), a shift by 33 days was revealed comparing data of 1961–1990 and 2051–2080 (REMO A1B). This means that hazel bloom would already start in average at January, 28th instead of March, 2nd, whereas considering REMO B1, a shift by 26 days was projected until 2051–2080. As WettReg assumes a more moderate temperature rise, the projected phenological shift of hazel bloom was not that intense: For WettReg A1B, a shift by 21 days was

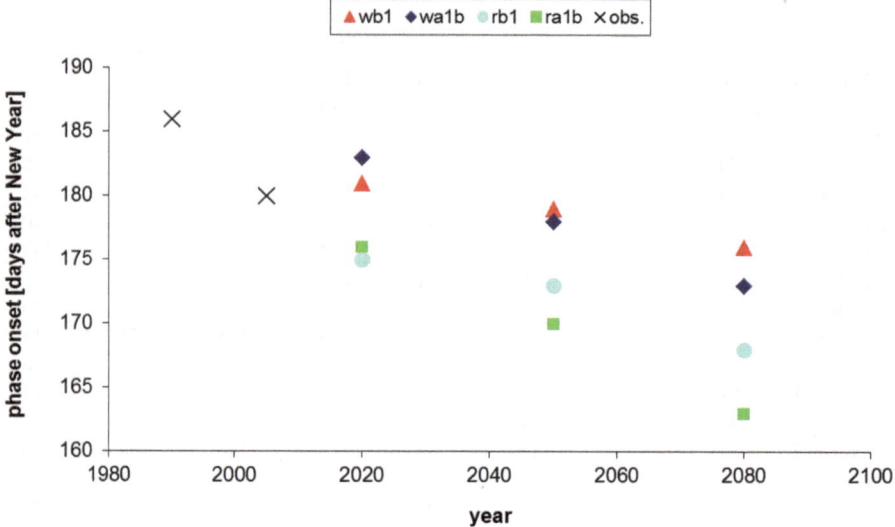

Lime bloom (phase 64) in ecoregion 12 (low mountain range)

Fig. 2.11 Mean onset of lime bloom in ecoregion 12 (low mountain range) calculated using observations (1961–2005) on phenology (*obs.* data provided by the German Weather Service) and projections on phenological development based on modelled air temperatures (2020–2080) (*w* WettReg, *r* REMO) for two different climate emission scenarios (B1, A1B)

shown whereas for WettReg B1 the shift amounts to 17 days. Future development of the beginning of hazel bloom resembles the development of the other investigated indicator phases, yet, there is a weakening of the shifts regarding the summer phases.

2.4.2 Spatially Discriminated Phenological Development

For detecting regional differences in phenological shifts, each grid on the phenological development (Sect. 2.4.1) was intersected with a map on German ecoregions (Sect. 2.2.3) and analysed statistically. Figure 2.11 depicts exemplarily some of the results derived from regionally differentiated analysis by example of the lime bloom in ecoregion 12 (low mountain ranges, Fig. 2.5).

The graph indicates the mean beginning of first lime bloom in ecoregion 12 based on observations of the German Weather Service and on both climate models (WettReg, REMO) regarding emission scenarios B1 and A1B (Sect. 2.2.2). Comparing the observation periods 1961–1990 and 1991–2005 with regard to ecoregion 12, there was a considerable shift of lime bloom of 6 days to the beginning of the year (black crosses). The observed advance of lime bloom should proceed assuming the REMO climate model. Considering emission scenario A1B (light green square), until 2080 the lime bloom should in average begin up to 17 days earlier (163 days

Table 2.5 Mean onset of phase 64 (flowering of large-leaved lime) (days after New Year) differentiated for the German ecoregions by Schröder and Schmidt 2001 for the observation periods 1961–1990, 1991–2005 and for the climate period 2051–2080 for the climate models WettReg and REMO and considering emission scenario B1 and A1B

Ecoregion	1961–1990	1991–2005	WettReg B1 2051–2080	WettReg A1B 2051–2080	REMO B1 2051–2080	REMO A1B 2051–2080
8	175	166	166	164	162	158
12	186	180	176	173	168	163
18	173	166	165	163	160	155
19	173	167	164	162	160	155
20	178	169	167	166	163	159
22	177	175	166	164	162	159
26	187	175	174	170	165	159
30	175	168	166	164	160	156
42	176	171	163	161	160	156
43	178	172	166	164	163	159
46	180	172	168	166	164	161
47	172	165	162	159	158	153
54	193	182	181	177	169	163
55	185	174	172	169	163	157
56	178	172	171	169	163	157
57	180	175	173	171	166	161
58	179	173	171	169	164	159
62	179	171	167	164	161	156
63	173	163	166	162	157	151
118	179	173	170	168	164	159
119	179	170	170	167	162	156
overall	178	171	168	166	162	157

after New Year) than in period 1991–2005. Considering WettReg climate model, projections of future phenological development were more moderate. In average, in 2080 lime bloom is expected to begin 173 days after New Year (dark blue square) which is 1 week earlier than in 1991–2005.

For most phenological phases and ecoregions these trends are comparable: projections based on REMO showed earlier beginnings compared to WettReg, projections for scenario B1 showed later beginnings than those for A1B. Only the magnitude of the differences varies from phase to phase and from period to period (Table 2.5).

2.5 Discussion and Conclusions

The projected shift in plant phenology is based on the regression between air temperature and plant phenology only. Thus, other driving factors like ground humidity or other soil conditions are not taken into account. However, fading out other relevant influences besides air temperature, it can be summarised that due to global warming a further shift of phenological phases to the beginning of the year has to be expected in future. The modelling approach presented in this study can help to identify areas at risk and to initiate adaptation strategies. Comparing both periods in the past (1961–1990, 1991–2005), the statistical analysis of all six investigated phases corroborated significant shifts towards the beginning of the year. This coincides with findings of Englert et al. (2008) who examined all 270 phases being observed by the German Weather Service. In spring, the onset of phenological phases was most distinct. Here, Pre-Spring proved to be the season mostly influenced by rising air temperature. On the other hand, in autumn 2 of 59 phases showed a moderate shift towards the end of the year.

The results presented here correspond to findings of, e.g., Ahas et al. (2002), Cyan et al. (2001), Fitter and Fitter (2002), Menzel et al. (2005), Schröder et al. (2006a, 2007), Stenseth et al. (2002), Theurillat and Guisan (2001) and Walther et al. (2002) at the regional to the European and global scale. The phenological shifts do not only affect the bio-geographical distribution of plants and the length of growing seasons (Dryer et al. 2006) but also fruit growing (Chmielewski et al. 2004; Wolfe et al. 2005) considering crop failures due to late frosts (Scheifinger et al. 2003) or pests (Hamilton et al. 2005) as well as bee-keeping and pollination service due to asynchronous flowering and pollinator's flight seasons (Parmesan et al. 2000, Penuelas and Filella 2001).

In this context, FISKA should be a useful tool for the early detection and assessment of possible impacts of global warming. Furthermore, it helps to initiate appropriate adaptation strategies by the federal authorities. The number of impact models and calculation kernels should be further enhanced to cover additional relevant cause-effect relationships including not only environmental concerns but also socio-economic issues like demographic change or effects on tourism. The information system is intended for integrating supplementary calculation kernels composed by other scientist. FISKA should be linked with other related information systems on climate impacts and opened for public awareness and discussion.

References

Ahas R, Aasa A, Menzel A, Fedotova VG, Scheifinger H (2002) Changes in European spring phenology. Int J Climatol 22:1727–1738

Breiman L, Friedman JH, Olshen RA, Stone CJ (1984) Classification and regression trees. Wadsworth International Group, Belmont

Chmielewski FM, Müller A, Bruns E (2004) Climate changes and trends in phenology of fruit trees and field crop in Germany, 1961–2000. Agric For Meteorol 121:69–78

Cyan DR, Kammerdiener SA, Dettinger MD, Caprio JM, Peterson DH (2001) Changes in the onset of spring in the Western United States. Bull Am Met Soc 82(3):399–415

Dryer LL, Esler KJ, Zietsman J (2006) Flowering phenology of South African Oxalis—possible indicator of climate change? S Afr J Bot 72(1):150–156

DWD (Deutscher Wetterdienst) (1991) Anleitung für die phänologischen Beobachter des Deutschen Wetterdienstes. Deutscher Wetterdienst, Offenbach

EEA (European Environmental Agency) (2008) Impacts of Europe's changing climate—2008 indicator-based assessment. Report No. 4/2008. http://www.eea.europa.eu/publications/eea_report_2008_4. Accessed April 2014

Englert C, Pesch R, Schmidt G, Schröder W (2008) Analysis of spatially and seasonally varying plant phenology in Germany. In: Car A, Griesebner G, Strobl J (eds) Geospatial crossroads @ GI_Forum '08: proceedings of the geoinformatics forum Salzburg. Wichmann, Heidelberg, pp 81–89

Fitter AH, Fitter RSR (2002) Rapid changes in flowering time in British plants. Science 296(5573):1689–1691

Hamilton JG, Dermody O, Aldea M, Zangerl AR, Rogers A, Berenbaum MR, DeLucia EH (2005) Anthropogenic changes in tropospheric composition increase susceptibility of soybean to insect herbivory. Environ Entomology 34(2):479–455

Hengl T, Heuvelink GBM, Rossiter DG (2007) About regression-kriging: from equations to case studies. Comput Geosci 33:1301–1315

IPCC (Intergovernmental Panel of Climate Change) (2007) Climate change 2007. Synthesis report. Geneva, 52 p

IPCC (International Panel on Climate Change) (2013) Acceptance of the actions at the twelfth session of Working Group I. Working Group I contribution to the IPCC Fifth Assessment Report (AR5), Climate change 2013: the physical science basis. Approved summary for policymakers. Thirty-six session of the IPCC, Stockholm, 26 September 2013, 36 p

Jacob D, Göttel H, Kotlarski S, Lorenz P, Sieck K (2008) Klimaauswirkungen und Anpassung in Deutschland—Phase 1: Erstellung regionaler Klimaszenarien für Deutschland. Forschungsbericht 204 41 138 UBA-FB 000969. Climate Change 11(08):1862–4359

Jaeger CC, Jaeger J (2011) Three views of two degrees. Reg Environ Change 11(Suppl 1):S15–S26

Johnston K, Ver Hoef JM, Krivoruchko K, Lucas N (2001) Using ArcGIS geostatistical analyst. ESRI, Redlands

Menzel A, Sparks TH, Estrella N, Eckhardt S (2005) 'SSW to NNE'—North Atlantic oscillation affects the progress of seasons across Europe. Glob Change Biol 11:909–918

Odeh IOA, McBratney AB, Chittleborough DJ (1995) Further results on prediction of soil properties from terrain attributes: heterotopic cokriging and regression-kriging. Geoderma 67(3–4):215–226

Parmesan C, Root T, Willig M (2000) Impacts of extreme weather and climate on terrestrial biota. B Am Meteorol Soc 81:443–450

Penuelas J, Filella I (2001) Phenology: responses to a warming world. Science 294:793–795

Roeckner E, Brokopf R, Esch M, Giorgetta M, Hagemann S, Kornblueh L, Manzini E, Schlese U, Schulzweida U (2006) Sensitivity of simulated climate to horizontal and vertical resolution in the ECHAM5 atmosphere model. J Climate 19(16):3771–3791

Romić MT, Hengl D, Romić D, Husnjak S (2007) Representing soil pollution by heavy metals using continuous limitation scores. Comput Geosci 33:1316–1326

Scheifinger H, Menzel A, Koch E, Peter CH (2003) Trends of spring time frost events and phenological dates in Central Europe. Theor Appl Climatol 74:41–51

Schmidt G (2002) Eine multivariat-statistisch abgeleitete ökologische Raumgliederung für Deutschland. dissertation.de, Berlin

Schmidt G, Holy M, Pesch R, Schröder W (2010) Changing plant phenology in Germany due to the effects of global warming. Int J Clim Change: Impacts Responses 2(2):73–84

Schröder W (2006) GIS, geostatistics, metadata banking, and tree-based models for data analysis and mapping in environmental monitoring and epidemiology. Int J Med Microbiol 296(40):23–36

Schröder W, Schmidt G (2001) Defining ecoregions as framework for the assessment of ecological monitoring networks in Germany by means of GIS and classification and regression trees (Cart). Gate to EHS 2001:1–9

Schröder W, Schmidt G, Hasenclever J (2006a) Geostatistical analysis of data on air temperature and plant phenology from Baden-Württemberg (Germany) as a basis for regional scaled models of climate change. Environ Monit Assess 130(1–3):27–43

Schröder W, Schmidt G, Hornsmann I (2006b) Landschaftsökologische Raumgliederung Deutschlands. In: Fränzle O, Müller F, Schröder W (eds) Handbuch der Umweltwissenschaften. Grundlagen und Anwendungen der Ökosystemforschung. Ecomed, München, pp 1–100 (Kap. V-1. 9, 17. Erg.Lfg)

Schröder W, Pesch R, Schmidt G (2007) Analysis of climate change affecting German forests by combination of meteorological and phenological data within a GIS environment. Scientific World 7(1):84–89

Schröder W, Pesch R, Schmidt G (2010) Projektion jahreszeitlicher Pflanzenentwicklungen im Klimawandel. In: Strobl J, Blaschke T, Griesebner G (eds) Angewandte Geoinformatik 2010. Beiträge zum 22. Agit-Symposium Salzburg. Wichmann, Heidelberg, pp 687–696

Spekat A, Enke W, Kreienkamp F (2006) Neuentwicklung von regional hoch aufgelösten Wetterlagen für Deutschland und Bereitstellung regionaler Klimaszenarien mit dem Regionalisierungsmodell WETTREG 2005 auf der Basis von globalen Klimasimulationen mit ECHAM5/MPI—OM T63L31 2010 bis 2100 für die SRES—Szenarien B1, A1B und A2. Projektbericht F + E-Vorhaben 204 41 138 Klimaauswirkungen und Anpassung in Deutschland—Phase 1: Erstellung regionaler Klimaszenarien für Deutschland. Potsdam, 94 p

Stenseth NC, Mysterud A, Ottersen G, Hurrell JW, Chan KS, Lima M (2002) Ecological effects of climate fluctuations. Science 297:1292–1296

Theurillat JP, Guisan A (2001) Potential impact of climate change on vegetation in the European Alps: a review. Clim Change 50(1–2):77–109

Walther GR, Post E, Convey P, Menzel A, Parmesan C, Beebee TJC, Fromentin JM, Hoegh-Guldberg O, Bairlein F (2002) Ecological responses to recent climate change. Nature 416:389–395

Wolfe DW, Schwartz MD, Lasko AN, Otsuki Y, Pool RM, Shaulis NJ (2005) Climate change and shifts in spring phenology of three horticultural woody perennials in northeastern USA. Int J Biometeorol 49:303–309

Chapter 3
Case Study 2: Phenological Trends in the Federal State of Hesse

Abstract In Hesse, a rise in air temperature of about 0.9 °C in average established comparing the climate reference period 1961–1990 and the period 1991–2009. This equals the trend in entire Germany, whilst in some regions even a temperature increase of up to 3 °C was measured. Until the end of the twenty-first century, air temperatures in Hesse are expected to rise between 3.2 and 3.7 °C compared to the reference period 1971–2000. This should also affect the onset and duration of phenological stages of plants implicating serious impacts for environment and economy. In order to deal with this issue, the research project "HeKlimPh" funded by the Hessian Competence Centre of Climate Change modelled spatiotemporal trends of plant phenology as being an indicator for climate change related biological effects. Accordingly, *case study 2* describes the coupled analysis of meteorological air temperature data and phenological data indicating different phenological seasons. The data comprise observations on 35 plant phenological phases of wild growing plants, fruits, crops and vines collected at about 6500 observation sites in Germany (553 in Hesse) between 1961–2005. Within a GIS environment, also estimations on the future phenological development for the periods 2031–2060 and 2071–2100 were performed. The projections were based on air temperature grids derived from four regional climate models considering the IPCC emission scenario A1B.

The statistical association between air temperatures and onset of phenological phases for the past periods 1961–1990, 1971–2000, and 1991–2009 was investigated by means of correlation analysis, considering auto-correlation, and regression analysis. For plant phases showing a significant and at least medium ($|r| \geq 0.5$) correlation between phase onset and air temperatures, phenological development was assessed by applying Regression Kriging. Future projections based on the regression models derived for the climate reference period 1971–2000. For each of the respective plant phases the according regression function was applied to air temperature grids depicting the estimated thermal development in the climate periods 2031–2060 and 2071–2100.

The calculations revealed that 31 out of 35 phases started earlier in the years 1991–2009 compared with 1961–1990. These shifts were more pronounced in Hesse (8 days) compared to the development in Germany (6 days). The onset of several plant phases was even more than 10 days earlier comparing the recent climate period and the prior one. Contrarily, plant phases indicating autumn and winter seasons tend to shift towards the end of the year yielding a prolongation of the vegetation period of up to 3 weeks. More than 70 % of the phases in each of the past periods were correlated with air temperatures by $r \leq -0.5$, more than 50 % even by $r \leq -0.7$. Projections gave reason to assume that the advances in phase onset to the beginning of the year should intensify in future: In many cases, shifts between 2071–2100 and 1961–1990 are expected to be at least twice as high as between 1991–2009 and 1961–1990.

Keywords Biomonitoring · Climate change impacts · Climate projections · Phenology mapping · Geostatistics

3.1   Background and Goals

Within the joint research cluster INKLIM-A maintained by the Hessian Compe-tence Centre of Climate Change[1], several research projects were initiated analysing direct and indirect impacts of climate change on environment and agriculture. This should help in developing adaptation strategies and mitigation measures to prevent from serious ecological and economic risks due to global warming. In order to cope with this issue, the research project "HeKlimPh" analysed and modelled spatial and temporal trends of plant phenology to support, for instance, land-use planning and management. Plant development in terms of phenological stages is a rather sensitive bio-meteorological response to environmental variation (Holopainen et al. 2013) and, thus, is ecologically relevant since changes in phenological development serve as both forcing and inhibiting ecological processes (Parmesan 2007) across spatial scales from individuals to landscapes (Rosenzweig et al. 2008) (Sect. 3.1). For ex-ample, shifts in phenology may affect crop yields due to frost damages or the spread of pests (Brown 2012; Inouye 2000; Li et al. 2014) and also address biodiversity issues concerning distribution and migration or disappearance of plant species.

   In Hesse, a rise in air temperature of about 0.9 °C in average established compar-ing the climate reference period 1961–1990 and the period 1991–2009. This equals the trend in entire Germany, whilst in some regions even a temperature increase of up to 3 °C was found (Sect. 2). Until the end of the twenty-first century, air tempera-ture in Hesse is expected to rise between 3.2 and 3.7 °C compared to the reference period 1971–2000 affecting the onset and duration of phenological stages of plants implicating serious impacts for environment and economy. Accordingly, the main goals of the "HeKlimPh" project were the analysis of the interdependence between air temperature rise and phenological development and, based on this, to generate maps on the onset of several indicator phases depicting the phenological develop-ment in Hesse in the past and in the future by means of Regression Kriging.

---

[1] http://klimawandel.hlug.de

3.2   Materials

Within a GIS environment, phenological data observed in Germany from 1961–2009 were compiled (Sect. 3.2.1) and statistically analysed with regard to the thermal development (Sect. 3.2.2) at the corresponding observation sites during the same climate periods (1961–1990, 1971–2000, 1991–2009). Based on the statistical findings, projections on the phenological development of the plant phases were calculated by use of four regional climate models considering the IPCC emission scenario A1B (IPCC 2007) for the periods 2031–2060 and 2071–2100 by means of Regression Kriging (Sect. 3.3).

3.2.1   Phenology Data

In Germany, systematic observation of plant phenology is maintained by the German Weather Service (DWD) and conducted by volunteers 2 or 3 times a week within a defined area with a radius of 5 km according to a guideline published by the DWD (1991) (Sect. 2.2.1). For *case study 2*, focusing on the federal state of Hesse in central Germany, a detailed analysis of the development of plants was performed by example of 35 plant phenological phases of wild growing plants (indicator and substitutional phases for the respective phenological season) and also for additional phases indicating different developmental stages of economically relevant fruits, crops and vines collected at about 6500 observation sites in Germany (553 in Hesse) between 1961–2009 (Table 3.1). Besides relevance for the phenological season, additional phases of wild growing plants were chosen completing missing vegetation layers (herbs, shrubs, trees).

For each observation site and each year, the phenological data contain the according onset of the particular phenophase in days after New Year. In an initial step, plausibility checks were performed in advance of the following statistical analyses to ensure reliable results. The phenological data were reviewed considering four criteria: (1) each observation site should cover at least 90 % of the respective long-term periods (1961–1990, 1971–2000, 1991–2005) (temporal representativeness); (2) sufficient and equal distribution of observations sites of a phase (spatial representativeness); (3) the onset at one site should not differ conspicuously from the average beginning at the surrounding observation sites during the same period (neighbourhood analysis); (4) the onset should not considerably begin earlier or later than compared to the average of all observations during the respective period (outlier analysis) (Fig. 3.1). If no reasonable explanation could be found regarding an identified conspicuous observation, the respective value was dropped. Consequently, it could happen that the whole site had to be excluded when criterion 1 was not met afterwards. Hence, the described quality control complemented the quality check routines conducted by the DWD.

For mapping the local phenological observation sites, according coordinates provided by the DWD were used to generate vector data (point layer) within a GIS.

Table 3.1 Phenological phases analysed in *case study 2* indicating different phenological seasons and representing different vegetation layers, complemented by relevant crops

Phase	Species	Phase onset	Vegetation layers	Relevance	Phenological season
Wild growing plants					
1	*Corylus avellana*	Flowering	Shrub/tree	Indicator	Pre-spring
2	*Galanthus nivalis*	Flowering	Herb	Substitute	Pre-spring
6	*Forsythia suspensa*	Flowering	Shrub	Indicator	First spring
52	*Ribes uva-crispa*	Unfolding of leaves	Shrub	Substitute	First spring
7	*Aesculus hippocastanum*	Unfolding of leaves	Tree		First spring
115	*Anemone nemorosa*	Flowering	Herb		First spring
62	*Malus domestica*	Flowering	Tree	Indicator	Full spring
13	*Quercus robur*	Unfolding of leaves	Tree	Substitute	Full spring
15	*Syringa vulgaris*	Flowering	Shrub		Full spring
19	*Alopecurus pratensis*	Full bloom	Herb		Full spring
18	*Sambucus nigra*	Flowering	Shrub	Indicator	Early summer
123	*Robinia pseudoacacia*	Flowering	Tree	Substitute	Early summer
20	*Dactylis glomerata*	Full bloom	Herb		Early summer
64	*Tilia platyphyllos*	Flowering	Tree	Indicator	Midsummer
100	*Ribes rubrum*	Fruit ripe for picking	Shrub	Substitute	Midsummer
109	*Malus domestica (early ripen.)*	Fruit ripe for picking	Tree	Indicator	Late summer
65	*Calluna vulgaris*	Flowering	Herb		Late summer
67	*Sambucus nigra*	First ripe fruits	Shrub	Indicator	Early autumn
177	*Rosa canina*	First ripe fruits	Shrub		Early autumn
72	*Quercus robur*	First ripe fruits	Tree	Indicator	Full autumn
68	*Aesculus hippocastanum*	First ripe fruits	Tree	Substitute	Full autumn
73	*Quercus robur*	Colouring of leaves	Tree	Indicator	Late autumn
226	*Quercus robur*	Leaf fall	Tree	Indicator	Winter
94	*Triticum aestivum*	Emergence	Herb	Indicator	Winter
Fruits					
54	*Prunus avium*	Flowering	Tree	Fruit growing	First spring
56	*Prunus cerasus*	Flowering	Tree		First spring
60	*Pyrus communis*	Flowering	Tree		First spring
102	*Prunus avium (early ripen)*	Fruit ripe for picking	Tree		Midsummer
103	*Prunus avium (late ripen)*	Fruit ripe for picking	Tree		Midsummer
104	*Prunus cerasus*	Fruit ripe for picking	Tree		Midsummer
107	*Pyrus communis (early ripen)*	Fruit ripe for picking	Tree		Early autumn
108	*Pyrus communis (late ripen)*	Fruit ripe for picking	Tree		Full autumn
Vines					
171	*Vine (Mueller-Thurgau)*	Sprouting	Shrub	Vine growing	First spring
172	*Vine (Mueller-Thurgau)*	Flowering	Shrub		Early summer
205	*Vine (Mueller-Thurgau)*	Grape gathering	Shrub		Full autumn

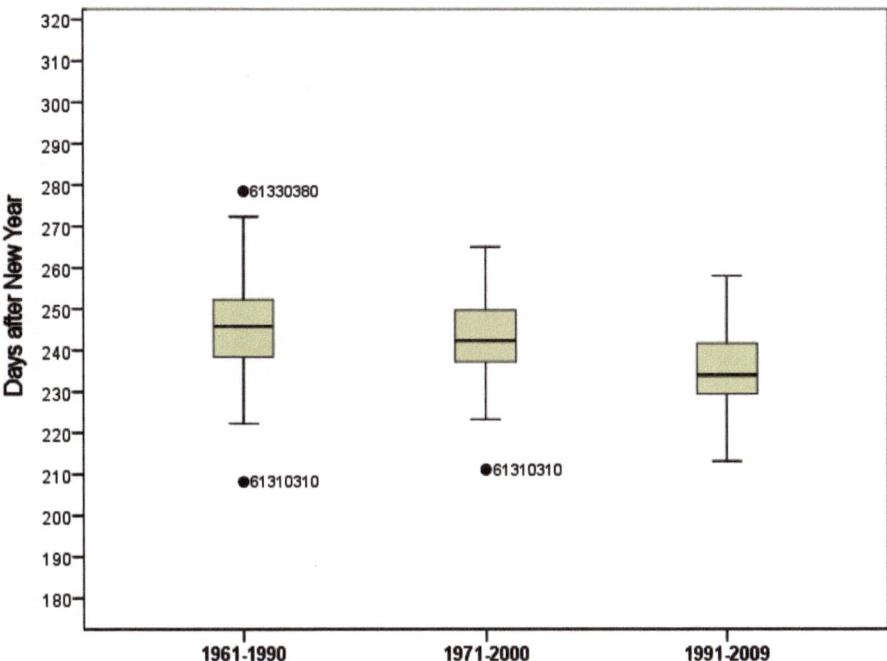

Fig. 3.1 Outlier (·) analysis by example of phase 67 (ripening of *Sambucus nigra*, *black* elder). For the observation periods 1961–1990 and 1971–2000 there were 3 outliers for two sites (indicated by the site no.) detected showing conspicuous values for the long-term phase onset

Subsequently, the point layers were intersected with the corresponding temperature grids (Sect. 3.2.2) to extract the respective air temperature values for each plant phase in each observation period at each location specifically.

3.2.2 Data on Air Temperatures

For each month from 1961–2009, the DWD provided air temperature grids with a spatial resolution of 1×1 km^2. These monthly means were aggregated to long-term monthly means for the periods 1961–1990, 1971–2000, and 1991–2009 (Figs. 3.3, 3.4, 3.5, 3.6, upper rows). Regarding air temperature development in the past, Fig. 3.2 shows the differences in long-term annual means between the periods 1971–2000 and 1961–1990 (left), 1991–2009 and 1971–2000 (centre), and 1991–2009 and 1961–1990 (right) regionally differentiated by the natural land units in Hesse. The differences in air temperature development in the two younger periods were in average twice as high (+0.6 °C) than compared to the two older reference periods (+0.3 °C). Accordingly, in sum there was a temperature rise in Hesse of about 0.9 °C comparing the periods 1991–2009 and 1961–1990 varying between 0.7 and 1.0 °C for the different Hessian land units.

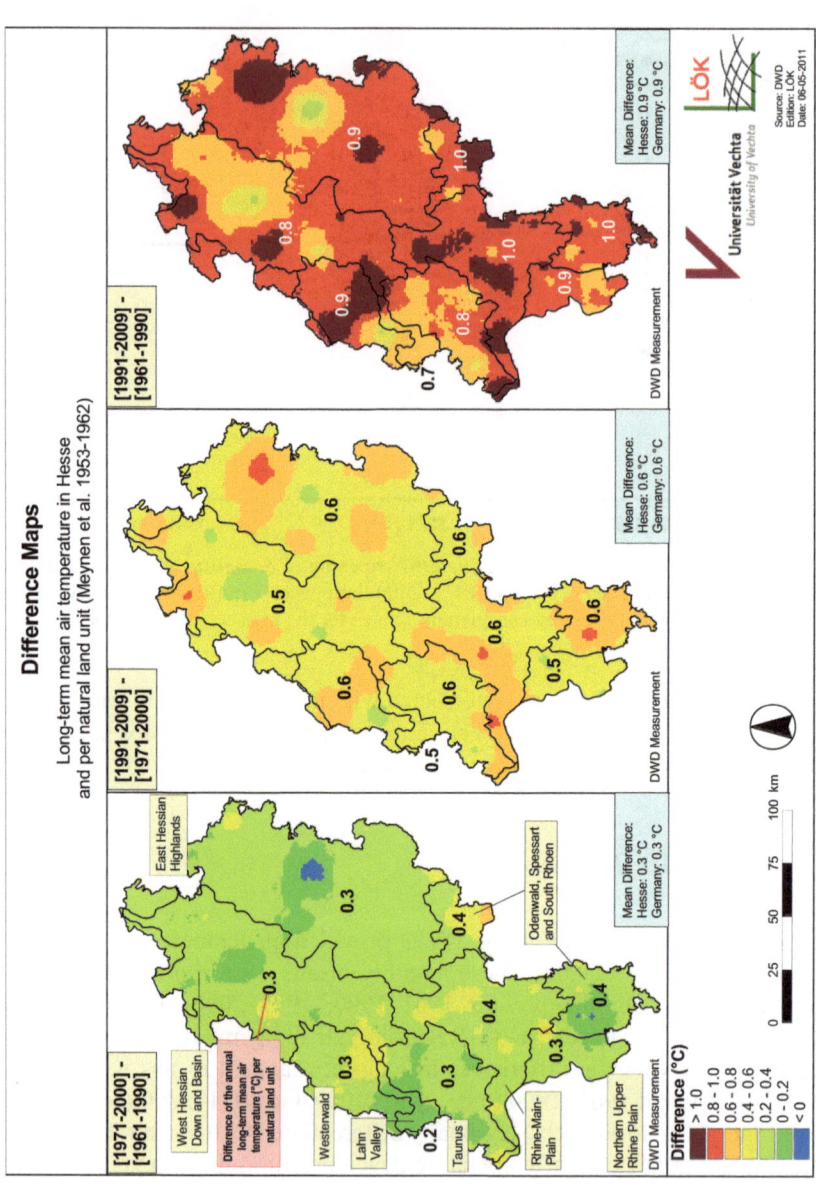

Fig. 3.2 Spatiotemporal patterns of air temperature development in Hessian natural land units depicted by maps indicating the differences between the long-term annual means for the periods 1971–2000 and 1961–1990 (*left*), 1991–2009 and 1971–2000 (*centre*), and 1991–2009 and 1961–1990 (*right*)

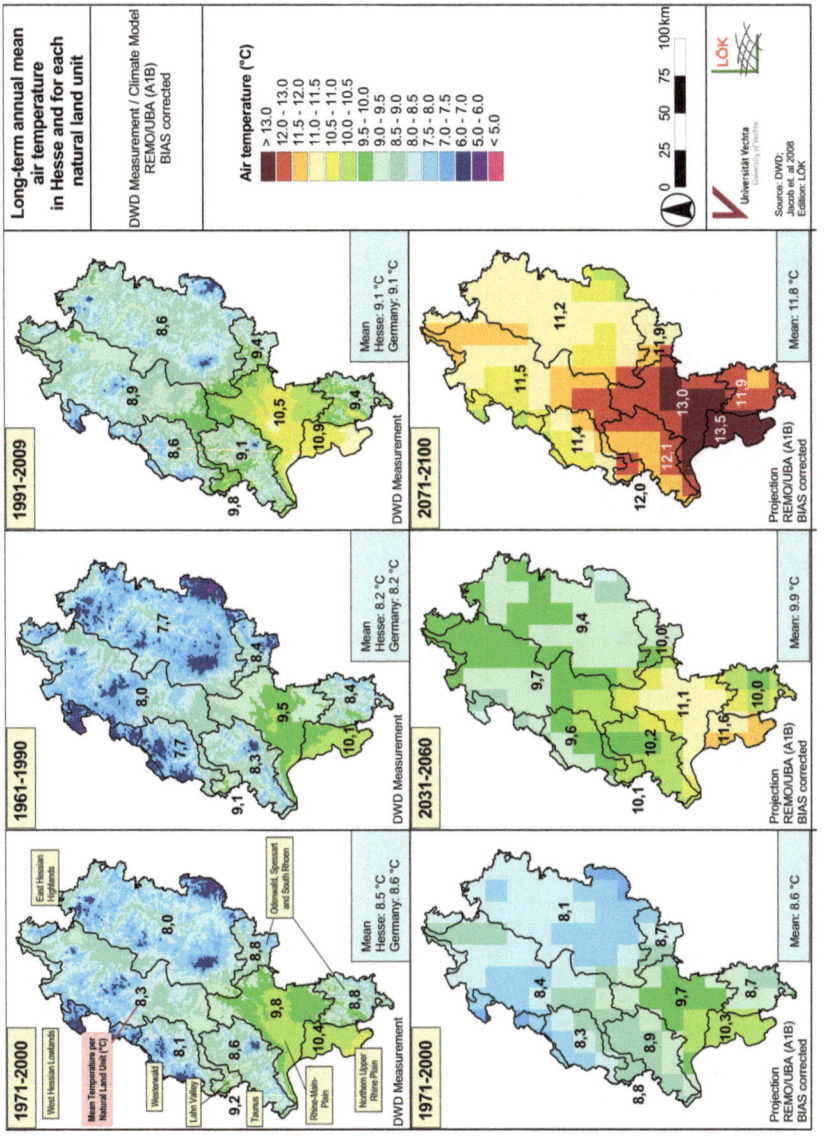

Fig. 3.3 Annual mean air temperatures in Hesse. *Above:* measurements from 1961 to 2009 by DWD; *below:* projections for 1971–2000, 2031–2060, and 2071–2100 for emission scenario A1B based on the REMO/UBA climate model

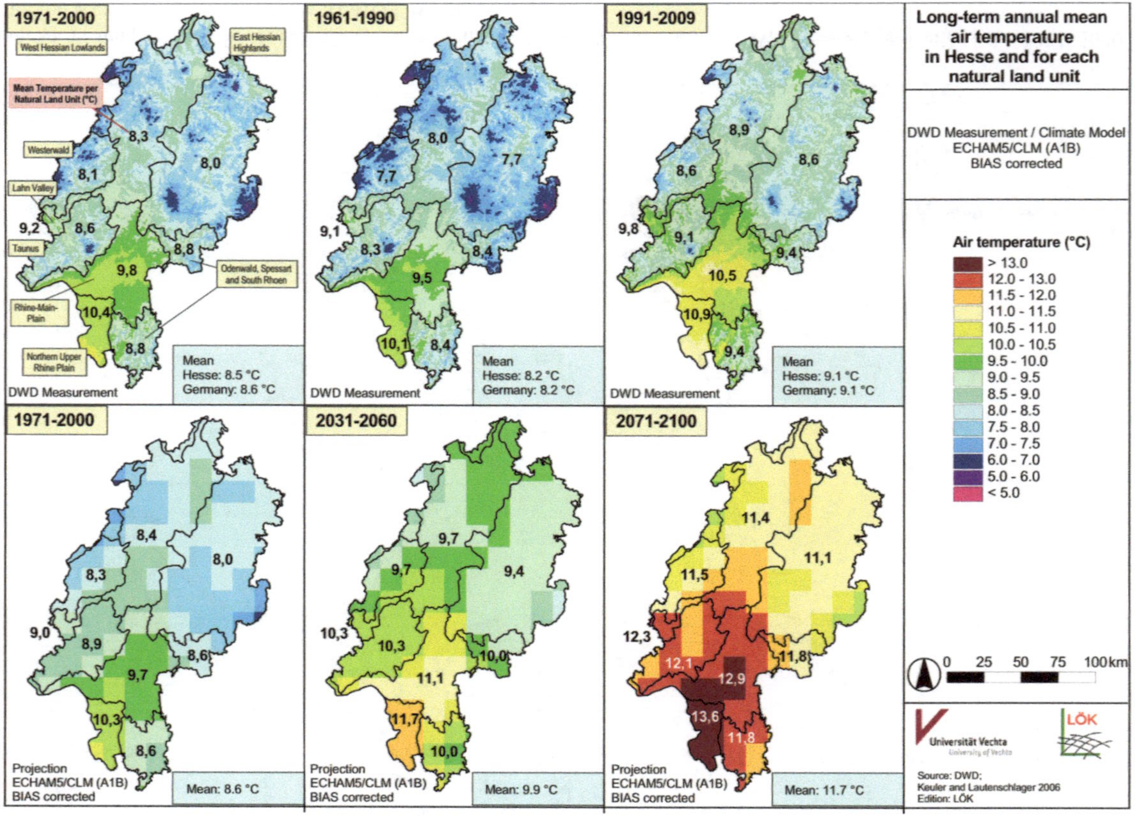

Fig. 3.4 Annual mean air temperatures in Hesse. *Above*: measurements from 1961 to 2009 by DWD; *below*: projections for 1971–2000, 2031–2060, and 2071–2100 for emission scenario A1B based on the ECHAM5/COSMO-CLM climate model

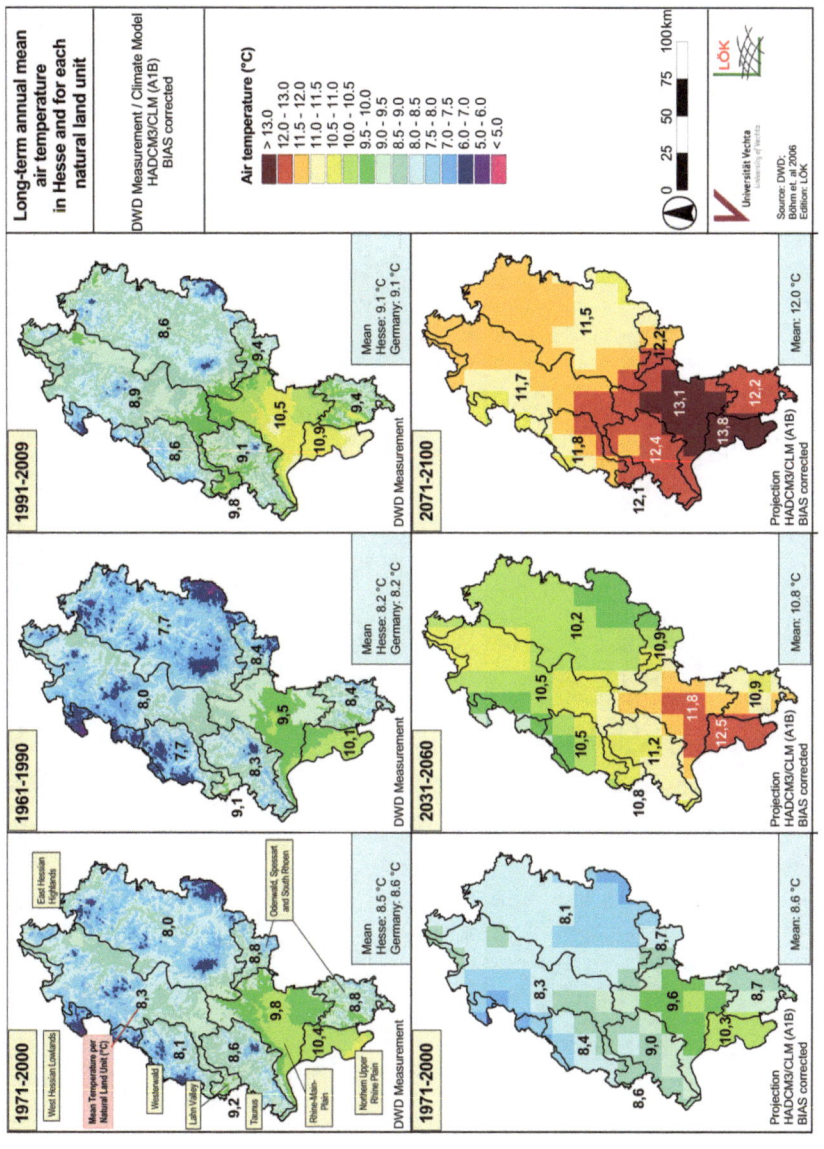

Fig. 3.5 Annual mean air temperatures in Hesse. *Above*: measurements from 1961 to 2009 by DWD; *below*: projections for 1971–2000, 2031–2060, and 2071–2100 for emission scenario A1B based on the HADCM3/COSMO-CLM climate model

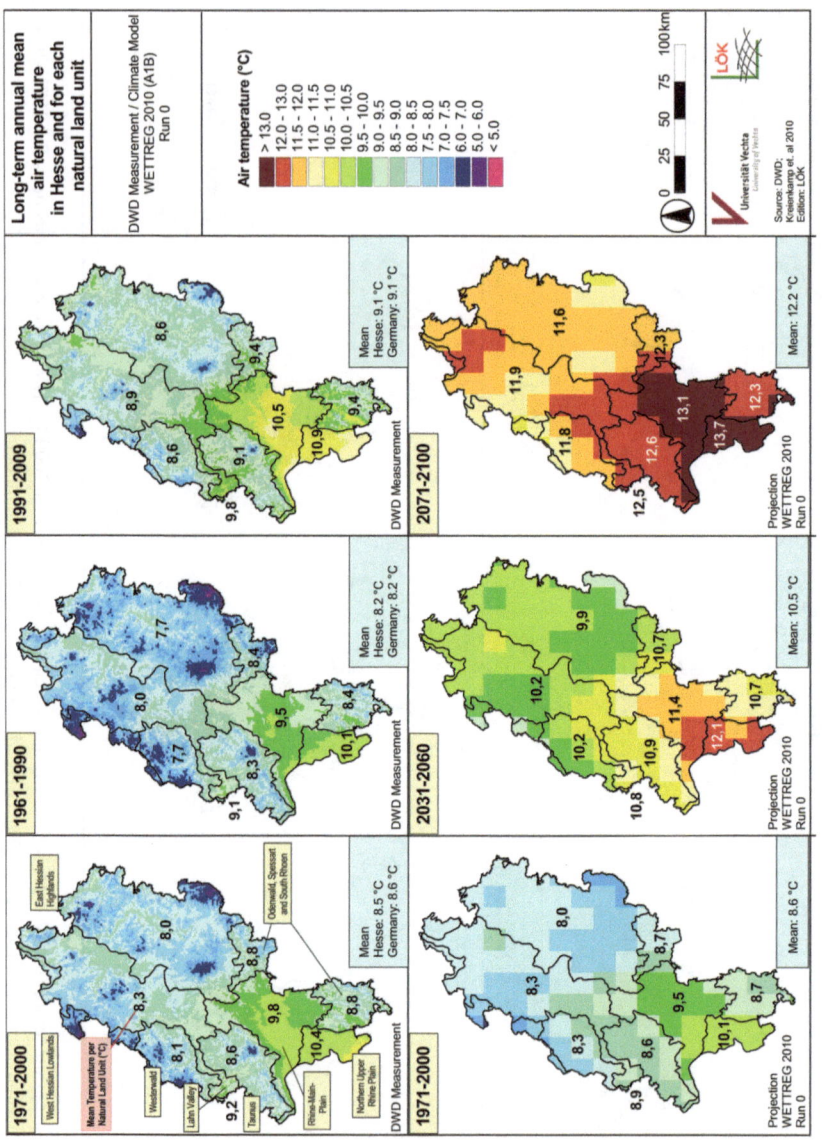

Fig. 3.6 Annual mean air temperatures in Hesse. *Above:* measurements from 1961 to 2009 by DWD; *below:* projections for 1971–2000, 2031–2060, and 2071–2100 for emission scenario A1B based on the WettReg2010 (run 0) climate model

The air temperature development in Hesse between 1961 and 2009 corresponds to the development in whole Germany (Sect. 2.2.2) showing an air temperature increase from in average 8.2 °C in the period 1961–1990 to 9.1 °C in the period 1991–2009. The maps reveal regional differences caused particularly by orographic patterns. Accordingly, lower regions are characterized by higher temperatures (e.g., Northern Upper Rhine Plain, Rhine Main Plain, Lahn Valley) whereas mountainous regions show rather colder temperatures (e.g. Westerwald, East Hessian Highlands).

For mapping the potential future phenological development in the periods 2031–2060 and 2071–2100 as well as for the reference period 1971–2000, results for the SRES A1B emission scenario (IPCC 2007) processed by four climate models were used within the INKLIM-A research consortium: REMO/UBA (Jacob et al. 2008) (Fig. 3.3), ECHAM5/COSMO-CLM (Keuler and Lautenschlager 2006) (Fig. 3.4), HADCM3/COSMO-CLM (Böhm et al. 2006) (Fig. 3.5), and WETTREG2010 (run 0, run 5) (Kreienkamp et al. 2010) (Fig. 3.6). The modelled long-term monthly means on projected air temperatures were provided as grids showing a spatial resolution of 20 km × 20 km^2. Figures 3.3–3.6 illustrate the observed spatial development of the long-term annual mean air temperature in Hesse between 1961 and 2009 (upper row) as well as the results of the different climate projections for the periods 1971–2000 (reference period), 2031–2060, and 2071–2100 (lower row).

Concerning the air temperature development between the reference period 1971–2000 and the period 2071–2100, air temperature increase is due to continue. However, the four climate models used for projecting the future phase onsets showed different characteristics (Fig. 3.7): Two of them projected rather moderate temperature rise until the end of the twenty-first century (ECHAM5/CLM, REMO/UBA) whereas the two others (HADCM3/CLM, WETTREG 2010) projected stronger shifts. Accordingly, mean annual air temperature rise in Hesse should range between 3.1 °C (ECHAM5/CLM) and 3.6 °C (WETTREG2010, run 0).

3.3 Methods

Following quality and plausibility checks (Sect. 3.2.1), the phenological observations were examined by descriptive statistics in the form of phenological clocks (Sect. 3.3.1) to detect trends in the past phenological development, and the interdependencies between meteorological and phenological data were investigated by correlation (Sect. 3.3.2) and regression analysis (Sect. 3.3.3). The mapping of phenological trends was realised by means of Regression Kriging (Sect. 3.3.4) based on the statistical associations between air temperature and phenological development.

3.3.1 Phenological Clocks

For describing the seasonal phenological development in Hesse in the course of the year so called 'phenological clocks' were prepared. In analogy to calendrical seasons (spring, summer, autumn, and winter), phenological seasons differentiate

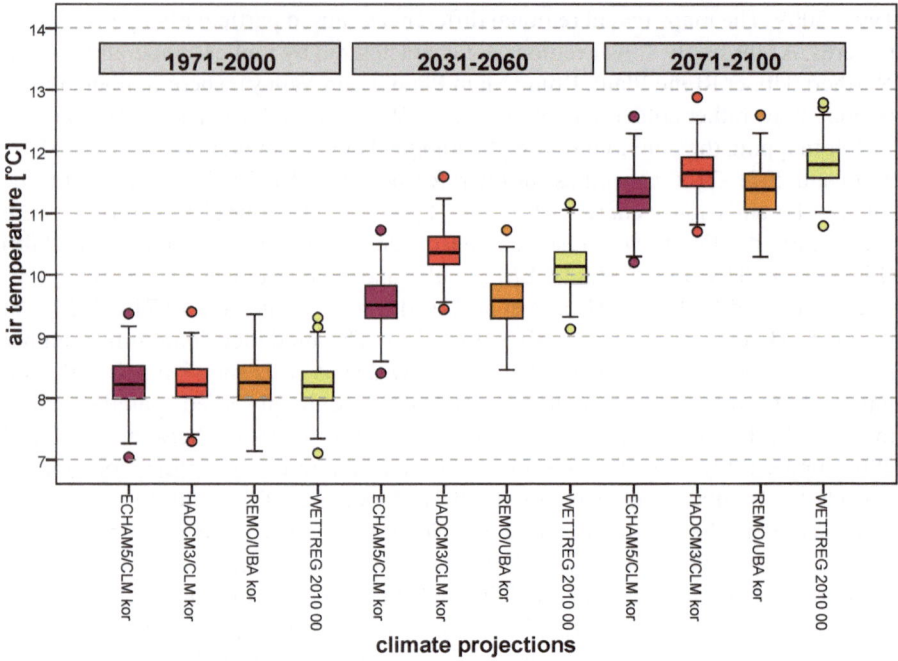

Fig. 3.7 Development of the projected mean annual air temperatures in Hesse as based on the climate models ECHAM5/CLM, HADCM3/CLM, REMO/UBA and WettReg2010 (run 0) regarding emission scenario A1B for the periods 1971–2000, 2031–2060, and 2071–2100

the vegetation period by the onset of so called 'indicator phases' dividing, for instance, the spring into three sub-seasons: Pre-Spring, First Spring, and Full Spring (Henniges et al. 2005). In case the indicator phases were not available, also substitutional phases may be considered. In the study at hand, the phenological clock contained three different rings within the clock: the innermost described the course of the developmental stages for the period 1991–2009, the middle one the development for the period 1971–2000 and the outer ring for the climate reference period 1961–1990. For a regionally discriminated analysis of the vegetation seasons (Sect. 3.4.2), phenological clocks for each land unit according to Meynen et al. (1953–1962) were assembled.

3.3.2   Correlation Analysis

The statistical association between air temperatures and the onset of phenological phases for the periods 1961–1990, 1971–2000, and 1991–2009 was investigated by means of bivariate correlation analysis (Pearson's product-moment correlation coefficients) and modelled by linear regression analysis (Sect. 3.3.3) considering also auto-correlation issues. In environmental systems, auto-correlation is a common phenomenon (Brown et al. 2006; Legendre 1993) which, in statistics, is defined as the similarity of, or correlation between, values of a process at neighbouring sites in time or space. Positive auto-correlation means that the individual observations contain information which is part of timely or spatial neighbouring observations. Thus, the effective sample size should be less than the total number of observations. Negative auto-correlation can have the opposite effect resulting in a larger effective sample compared to the realized sample. Hence, auto-correlation may have several implications for, e.g., statistical inference testing and regression analysis (Dale and Fortin 2009). Considering parametric statistics such as Pearson's correlation coefficients, for instance, positive spatial auto-correlation may enhance type I errors causing statistical significance even when there is no (Fortin and Payette 2001). Thus, in this investigation, spatial auto-correlation was considered for the correlation analyses according to Dutilleul (1993).

3.3.3 Regression Analysis

Regression analysis between air temperatures and phenological onset considered those months in the course of the year showing the strongest correlation between air temperatures and the respective phenological onset instead of only using annual or monthly mean air temperatures. Accordingly, for each phenophase and period (1961–1990, 1971–2000, 1991–2009) long-term mean temperatures were calculated for each distinct sequence of these months (e.g., averaged temperatures from March to June for the period 1971–2000). This approach should take the aspect of dormancy into account, as (1) temperatures of up to several months in advance of the respective phase onset affect the date of this event as well as (2) temperature conditions close to the eventual date of occurence (Schnelle 1955). The resulting r-values were classified as follows: $|r| < 0.20$: very low correlation, $0.20 \leq |r| \leq 0.49$: low, $0.5 \leq |r| \leq 0.69$: medium, $0.70 \leq |r| \leq 0.89$: high, $|r| \geq 0.9$: very high correlation (Hagl 2008).

3.3.4 Regression Kriging

For those phases showing a significant and at least medium correlation ($|r| \geq 0.5$), phenological maps for each period (1961–1990, 1971–2000, 1991–2009, 2031–2060, 2071–2100) were calculated in a GIS by Regression Kriging (Odeh et al. 1995; Zirlewagen et al. 2007). The regression equation derived for each phase and period (Sect. 3.3.3) was thereby applied to the long-term mean temperature grids of Germany (Sect. 3.2.2) to calculate surface maps for the respective phenophase. The resulting regression maps were added by Kriging maps depicting the spatial structure of the residuals of the regression models based on auto-correlation functions determined by use of variogram analysis (Hengl et al. 2007; Romić et al. 2007).

Future projections based on the regression models derived for the climate reference period 1971–2000. For each of the plant phases, the according regression function was applied to air temperature grids depicting the estimated thermal development in the climate periods 2031–2060 and 2071–2100 with regard to the four climate models used in INKLIM-A (Sect. 3.2.2). For validation issues, also for the climate reference period 1971–2000 map projections were calculated for each of the four climate models to compare the results with the maps derived from observed air temperature for the particular period (Sect. 3.4.1). For plant phases showing a significant and at least medium ($|r| \geq 0.5$) correlation between phase onset and air temperatures, phenological development was assessed by applying Regression Kriging. Additionally, the past and potential future phenological development was spatially differentiated by intersection of the respective phenological maps with a map on the natural land units (Meynen et al. 1953–1962) of Hesse (Sect. 3.4.2).

3.4 Results

3.4.1 Phenological Development in the Past

To illustrate phenological trends from 1961–2009, the long-term course of the onset of each of the 35 investigated phenophases (Sect. 3.2.1) was described by corresponding graphs, exemplarily depicted in Fig. 3.8 for phase 67 (first ripe fruits of *Sambucus nigra*, black elder). It reveals similar long-term trends of the phase onset for Hesse and Germany, as the onset shifts towards the beginning of the year comparing the periods 1961–1990 and 1991–2009 (1991–2005 for Germany). However, the regional shift in Hesse (−9 days) is stronger than compared to the national level (−7 days). In the long-term mean of the period 1961–1990, fruit ripeness of black elder started in Hesse on September 1st (244th day of the year), whereas the mean beginning of phase 67 in the period 1991–2009 was on August 23rd (235th day of the year).

Furthermore, Fig. 3.8 shows that the phase onset is subject to distinct annual variation, however, a high congruence between the respective annual phase onsets on national and regional scale can be stated. For instance in 1983, fruit ripening of black elder in Hesse began 19 days earlier than in 1984 (in Germany 16 days earlier). This phenological variation went along with air temperature development, as 1983 was characterized by rather mild temperatures, whereas 1984 was a distinct cooler year. This example underlines the importance of taking annual variation within long-term periods into account to draw reasonable conclusions.

Considering the long-term development of all examined phases, nearly all of them (31 of 35 phases) showed a shift in phenological onset to the beginning of the year between the periods 1961–1990 and 1991–2009 (Fig. 3.9). As it holds true for phase 67 previously described, in average of all 35 phases, shifts in Hesse were even stronger (about 8 days) than in Germany (about 6 days). Many phases even showed a shift of more than 10 days. The strongest shifts were detected for phases in spring and early summer. For instance, hazel bloom (*Corylus avellana*) in Hesse began

Beginning of fruit ripening of black elder (*Sambuccus nigra*), phase 67
Medians of phase onset in each year from 1961-2009

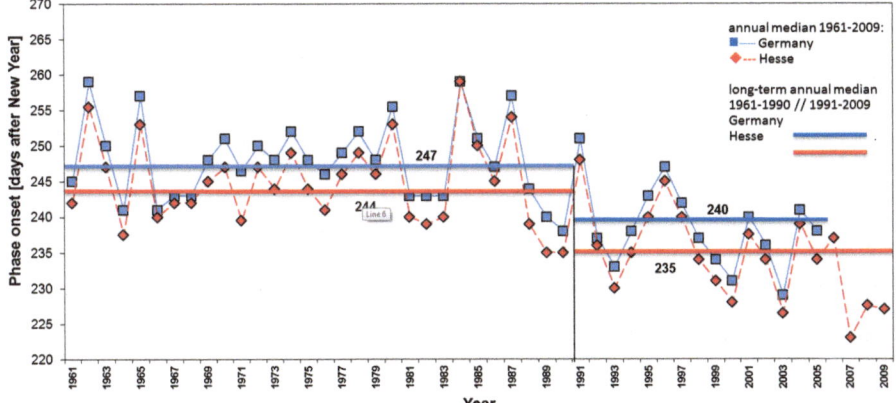

Fig. 3.8 Long-term annual course of the beginning of ripening of *Sambucus nigra* spatially differentiated for Germany and the federal state of Hesse and temporally differentiated for the periods 1961–1990 and 1991–2009

12 days earlier in the period 1991–2009 compared to the period 1961–1990. In the further course of the year, some phases—especially in late summer and autumn—showed weaker shifts. However, fruit ripeness of the black elder as illustrated above constitutes an exception within the late season's phenophases of the year. At the end of the phenological year in late autumn and winter, some phases (e.g., decolouring and fall of leaves of the pedunculate oak) even showed a reverse shift towards the end of the year (Fig. 3.9). As a result, a prolongation of the vegetation period was observed which amounted up to 10 days in some natural land units in Hesse. Some land units even showed prolongations of almost 3 weeks (Sect. 3.4.2).

Besides the phase specific advances to the beginning of the year, Fig. 3.9 reveals temporal variations of phase shifts as well: Shifts between the more recent periods 1991–2009 and 1971–2000 are stronger than between the periods 1971–2000 and 1961–1990. This development corresponds with air temperature rise: Warming in Hesse between the two more recent periods were twice as strong as between the two older periods (Sect. 3.2.2).

3.4.2 Spatially Discriminated Phenological Development

For detecting regional differences in phenological shifts, phenological clocks were generated as described in Sect. 3.3.1. Figure 3.10 illustrates the regionally differentiated findings for each of the natural land units in Hesse.

This regionally differentiated analysis revealed that (1) in the lowlands, e.g., in the Northern Upper Rhine Valley (unit 220), phase onset in average occurred much earlier (up to 3 weeks) than in the Eastern Hessian Highlands (unit 350). (2)

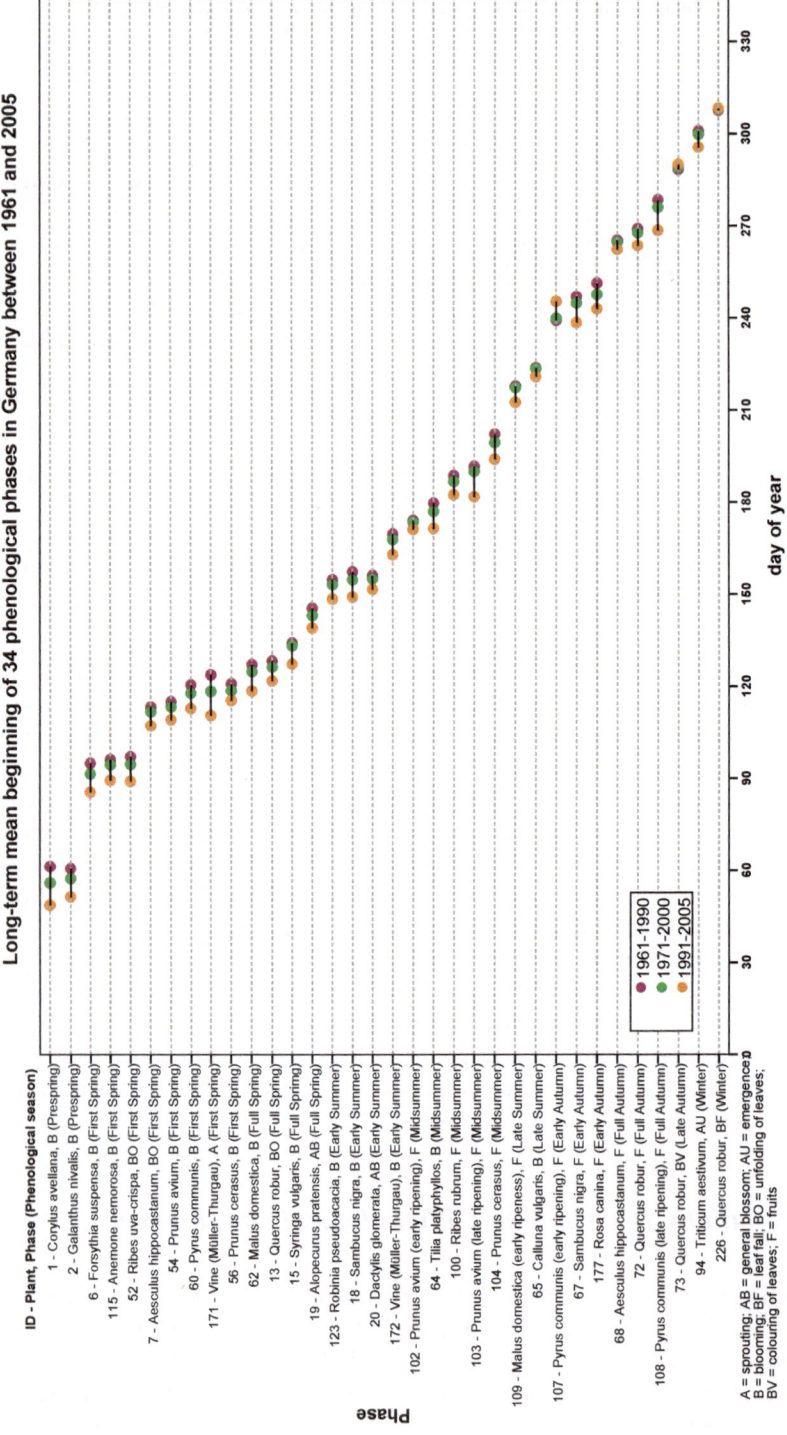

Fig. 3.9 Shifts in phenological onset in Hesse depicted for each phase investigated and temporally differentiated for the periods 1961–1990 (*red dots*), 1971–2000 (*green dots*), and 1991–2009 (*orange dots*)

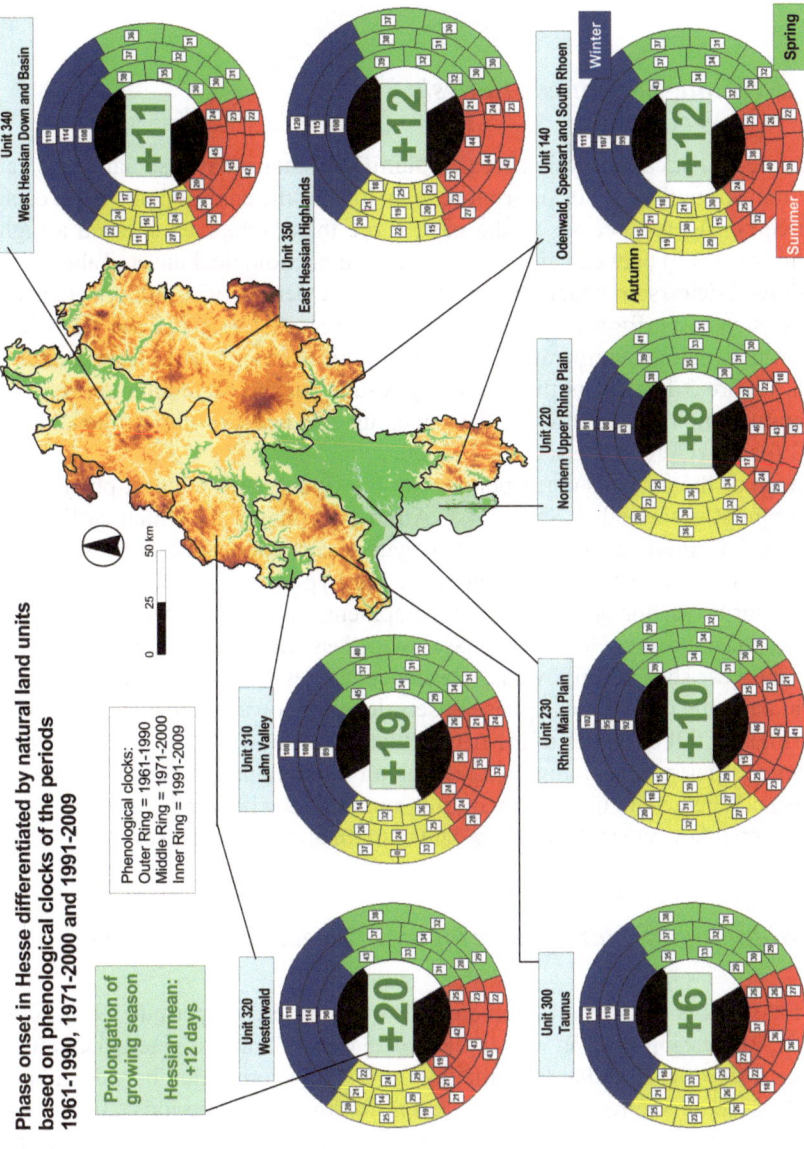

Fig. 3.10 Phenological clocks for Hesse for the periods 1961–1990 (*inner* ring), 1971–2000 (*middle* ring); and 1991–2009 (*outer* ring) spatially differentiated for the natural land units. The values in the centre of the rings indicate the differences in the length (*prolongation*) of the vegetation periods 1961–1990 and 1991–2009 in Hesse

In summary, most mountainous regions in Hesse were affected by stronger shifts of phase onset resulting in a more distinct prolongation of the vegetation periods (up to 20 days for the Westerwald, unit 320) whereas the river valleys in the south of Hesse were affected more moderate (for example, only 8 days for the Northern Upper Rhine Valley).

3.4.3   Bivariate-Statistical Analysis

The bivariate statistical analysis revealed significant statistical associations of at least medium strength ($r \leq -0.5$) for more than 70% of the analysed phases in each of the three considered periods in the past. More than 50% even showed a high correlation ($r \leq -0.7$) between air temperatures and phenological onset (Table 3.2).

The derived regression equations for the reference period 1971–2000 were used in the next step to perform Regression Kriging (Sect. 3.4.4) given the statistical association between air temperature and phenological onset was at least medium ($r \leq -0.5$). Figure 3.11 depicts the according results for the regression analysis of air temperature and the beginning of fruit ripening of *Sambucus nigra* (black elder, phase 67) for the reference period 1971–2000.

The result of the analysis corresponds with the findings for the past phenological development described in Sect. 3.4.1: Almost all phases for which earlier beginnings were figured out showed high negative correlation coefficients. These findings corroborate, spatially differentiated, the hypothesis that air temperature is a significant driver for phenological development: high air temperatures induce early phenophase onsets. However, phases with less intense shifts in the further course of the year showed only weak interdependencies with air temperatures, especially in autumn. Two of those phase onsets delaying towards the end of the year (decolouring and fall of leaves of the penduculate oak) showed even positive weak correlation coefficients. This implies that high air temperatures in autumn have reverse effects: Whereas high air temperatures stimulate the beginning of spring and summer phases, they retard the onset of late autumn and winter phases.

3.4.4   Mapping of Plant Phenological Development in Hesse

Based on the results of the regression analysis of the reference periods 1971–2000 (Table 3.2), Regression Kriging was applied to 23 of the 35 investigated plant phases showing a significant and at least medium ($r \leq -0.5$) correlation between phase onset and air temperatures. Respective maps for the future periods 2031–2060 and 2071–2100 were calculated applying regression equations of the observation period 1971–2000 to the temperature grids derived from the four climate projections (Sect. 3.3.4). For validation of the respective climate model, also maps for the reference 1971–2000 were calculated by use of air temperature projections. Figures 3.12, 3.13, 3.14, 3.15, 3.16 depict the results of the Regression Kriging

Table 3.2 Results of the bivariate analyses of air temperatures and phenological onset for all 35 phases investigated in the respective climate periods 1961–1990, 1971–2000, and 1991–2009; R^2 = coefficient of determination, r = correlation coefficient (Pearson)

Phase	1961–1990		1991–2009		1971–2000		Regression equation 1971–2000
	R^2	R	R^2	r	R^2	R	
1	0.62	−0.79	0.50	−0.71	0.62	−0.79	y=−0.9015x+82.086
2	0.56	−0.75	0.41	−0.64	0.54	−0.73	y=−0.6505x+75.802
6	0.72	−0.85	0.68	−0.82	0.77	−0.88	y=−0.8158x+134.35
52	0.49	−0.70	0.38	−0.62	0.48	−0.69	y=−0.5742x+141.29
7	0.53	−0.73	0.52	−0.72	0.59	−0.77	y=−0.5961x+149.34
115	0.27	−0.52	0.27	−0.51	0.37	−0.61	y=−0.4296x+128.76
62	0.68	−0.82	0.68	−0.83	0.76	−0.87	y=−0.6356x+188.82
13	0.53	−0.73	0.49	−0.70	0.59	−0.77	y=−0.5858x+196.05
15	0.69	−0.83	0.69	−0.83	0.73	−0.86	y=−0.6205x+206.59
19	0.21	−0.46	0.20	−0.45	0.28	−0.53	y=−0.4323x+194.16
18	0.56	−0.75	0.55	−0.74	0.59	−0.77	y=−0.5885x+232.59
123	0.36	−0.60	0.49	−0.70	0.54	−0.73	y=−0.5707x+212.07
20	0.31	−0.56	0.39	−0.63	0.42	−0.65	y=−0.4794x+221.58
64	0.50	−0.71	0.40	−0.63	0.58	−0.76	y=−0.6543x+263.53
100	0.54	−0.73	0.37	−0.61	0.53	−0.73	y=−0.6629x+290.46
109	0.57	−0.76	0.38	−0.62	0.55	−0.74	y=−0.7761x+341.31
65	0.01	−0.11	0.00	−0.06[a]	0.01	−0.07[a]	y=−0.0536x+231.98
67	0.31	−0.55	0.32	−0.57	0.31	−0.56	y=−0.6185x+343.74
177	0.15	−0.39	0.11	−0.33	0.14	−0.38	y=−0.6215x+335.26
72	0.09	−0.30	0.12	−0.34	0.12	−0.34	y=−0.3614x+315.3
68	0.15	−0.38	0.19	−0.43	0.16	−0.40	y=−0.2841x+301.81
73	0.01	0.12	0.02	0.13	0.02	0.14	y=0.1138x+279.03
94	0.01	0.12	0.05	0.23	0.03	0.16	y=0.172x+284.73
226	–	–	0.02	0.15	–	–	–
54	0.68	−0.82	0.70	−0.84	0.73	−0.86	y=−0.6022x+151.9
56	0.69	−0.83	0.66	−0.81	0.73	−0.85	y=−0.6036x+179.69
60	0.71	−0.84	0.71	−0.84	0.77	−0.88	y=−0.7225x+177.63
102	0.47	−0.69	0.46	−0.68	0.58	−0.76	y=−0.7682x+291.62
103	0.42	−0.65	0.36	−0.60	0.55	−0.74	y=−0.5999x+285.02
104	0.40	−0.63	0.34	−0.58	0.40	−0.63	y=−0.575x+295.94
107	0.26	−0.51	0.26	−0.51	0.40	−0.63	y=−0.8363x+375.93
108	0.09	−0.30	0.08	−0.28	0.15	−0.39	y=−0.3998x+320.82
171	0.19	−0.44	0.38	−0.62	0.26	−0.51	y=−0.5107x+155.9
172	0.39	−0.62	0.73	−0.85	0.60	−0.78	y=−0.6548x+253.1
205	–	–	0.13	−0.36	0.61	−0.78	y=−1.1315x+426.6

[a] correlation was not statistically significant

approach for the beginning of fruit ripeness of *Sambucus nigra* (black elder) being the indicator phase for early autumn (phase 67). The upper rows of Figs. 3.12–3.16 illustrate the observed long-term mean phase onset of the past periods 1961–1990, 1971–2000, and 1991–2009. The lower rows depict the future phenological development of the onset of phase 67 based on projected air temperature data according to the respective climate model.

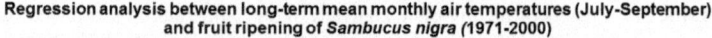

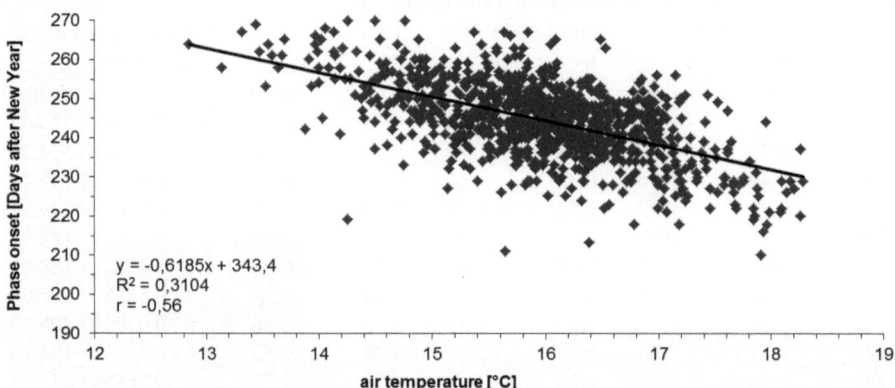

Fig. 3.11 Regression analysis for the correlation between long-term mean monthly air temperatures (July–September) and the beginning of fruit ripening of *Sambucus nigra* (black elder, phase 67) for the reference period 1971–2000

In accordance with the findings in Sect. 3.4.1, fruit ripeness of the black elder started considerably earlier in the long-term mean of the period 1991–2009 compared to the periods 1961–1990. With regard to the surface estimations, fruit ripeness occurred in average 237 days after New Year (August 25) instead of 246 days (September 3) resulting in an advance of nine days. In comparison to the mean values derived from the grids, the local observations of the German Weather Service resulted in an averaged shift of 11 days from September 3 (246 days after New Year) during the period 1961–1990 to August 23 during the period 1991–2009 (235 days).

As could be expected, topographical patterns were traced by the phenological surface maps: Lower regions indicating rather warm temperatures were characterized by early phase onset (e.g., Northern Upper Rhine Plain in the southwest). In comparison, mountainous regions (e.g., the East Hessian Highlands in the northeast) showed late phase onsets. In Figs. 3.12–3.16, for each natural land unit the according long-term mean phase onset was calculated for a spatially differentiated view. For instance, in average of the period 1961–1990 fruit ripeness of the black elder (phase 67) started 232 days after New Year in the Northern Upper Rhine Plain and nearly 3 weeks later in the East Hessian Highlands (252 days after New Year).

In compliance with the findings of Sect. 3.4.1 and corresponding with the development of air temperature rise in Hesse between the 1961 and 2009, showing a far stronger warming between the younger periods 1971–2000 and 1991–2009 compared to the older periods 1961–1990 and 1971–2000 (Sect. 3.2.2), the advance of the onset of fruit ripeness of *Sambucus nigra* was far more intense between the younger periods 1971–2000 and 1991–2009 (−7 days) as compared to the shifts between the older periods 1961–1990 and 1971–2000 (−2 days) (Fig. 3.17). Comparing the two older periods, a maximum shift of 5 days towards

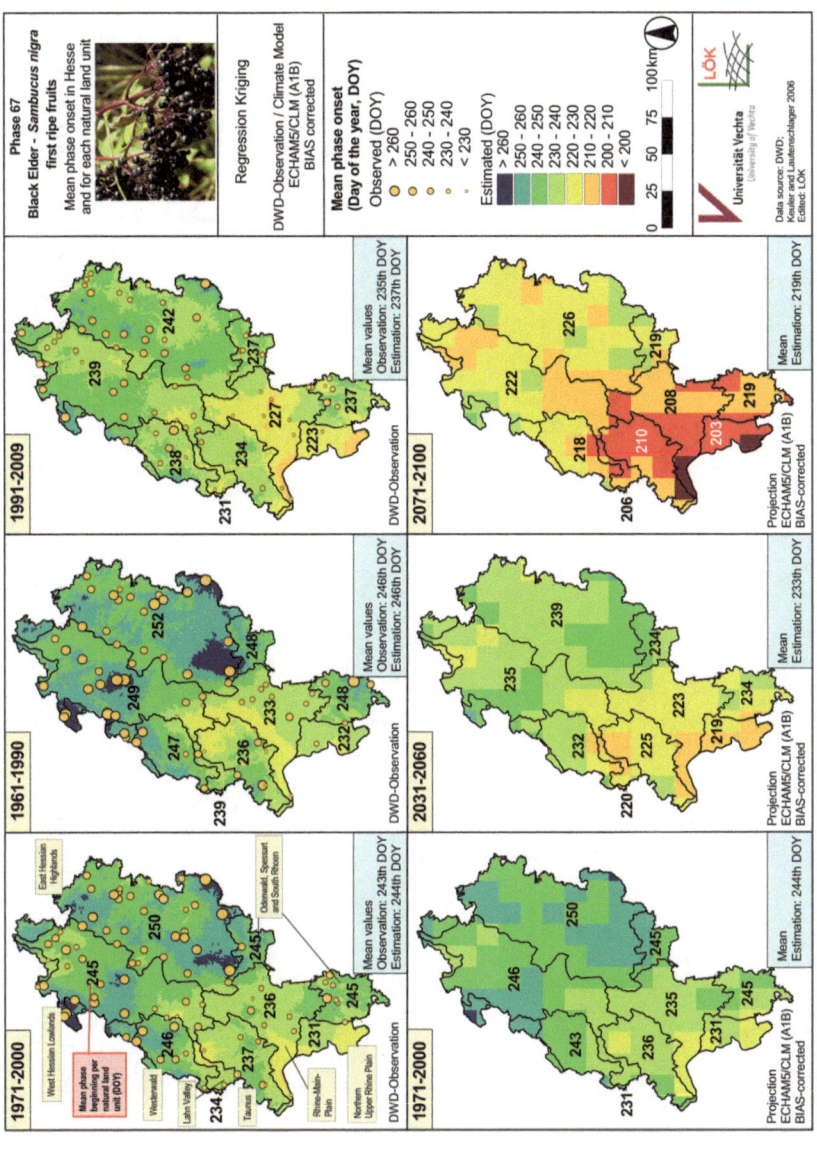

Fig. 3.12 Surface maps derived by Regression Kriging for the beginning of fruit ripening of *Sambucus nigra* (*black elder*) in Hesse based on observed (1961–1990, 1971–2000, 1991–2009) and projected air temperature grids (1971–2000, 2031–2060, 2071–2100) derived from ECHAM5/CLM climate model, emission scenario A1B

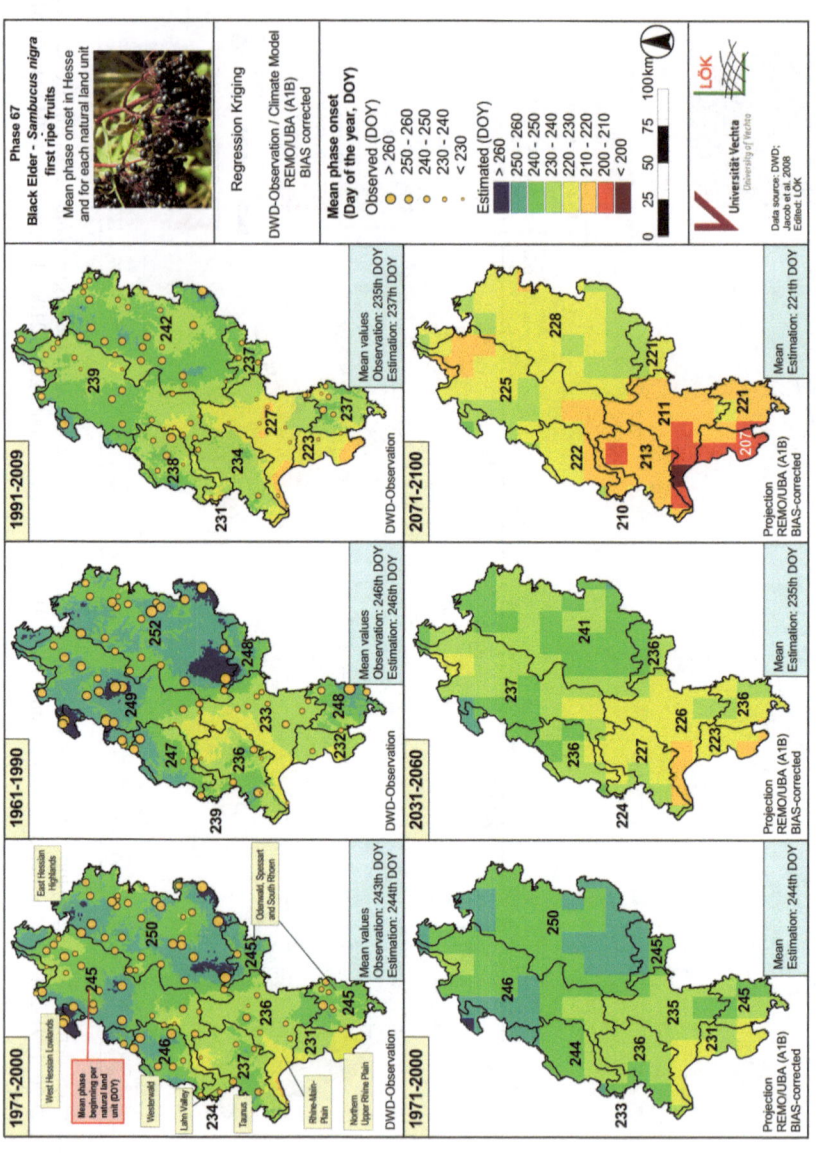

Fig. 3.13 Surface maps derived by Regression Kriging for the beginning of fruit ripening of *Sambucus nigra* (*black elder*) in Hesse based on observed (1961–1990, 1971–2000, 1991–2009) and projected air temperature grids (1971–2000, 2031–2060, 2071–2100) derived from REMO/UBA climate model, emission scenario A1B

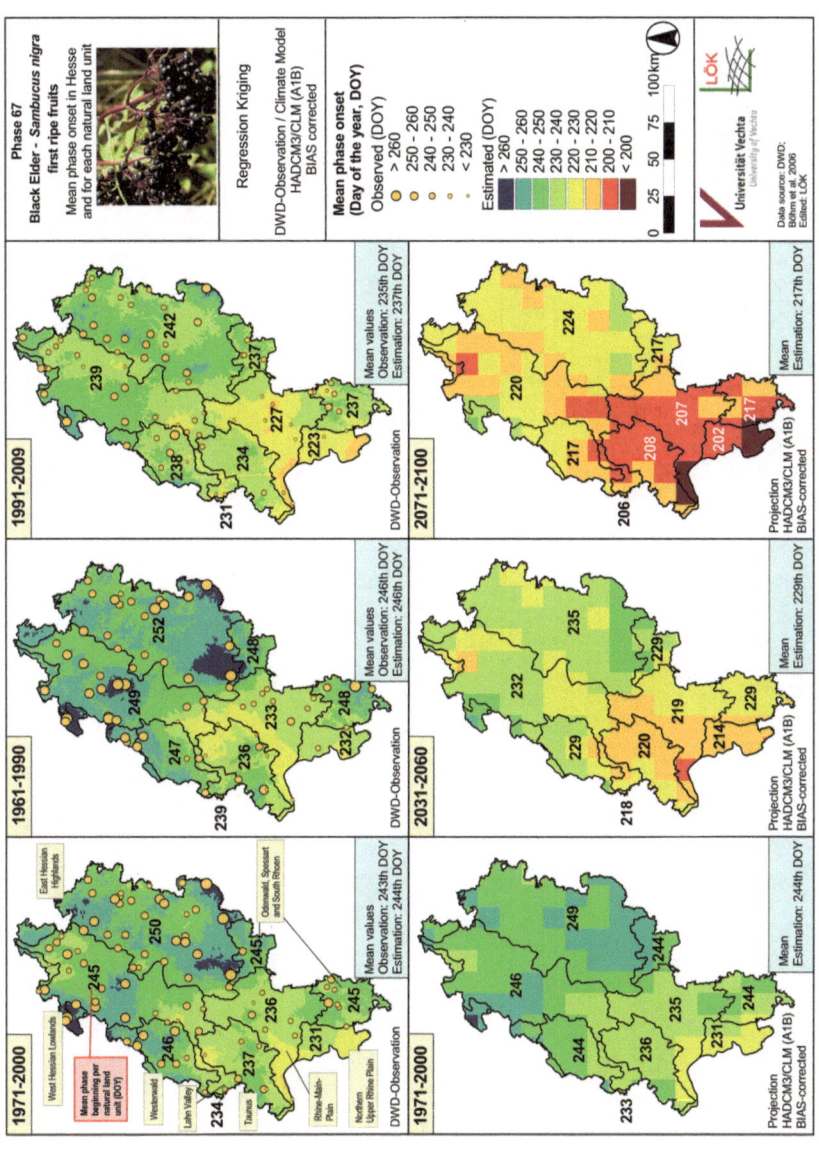

Fig. 3.14 Surface maps derived by Regression Kriging for the beginning of fruit ripening of *Sambucus nigra* (*black* elder) in Hesse based on observed (1961–1990, 1971–2000, 1991–2009) and projected air temperature grids (1971–2000, 2031–2060, 2071–2100) derived from HADCM3/CLM climate model, emission scenario A1B

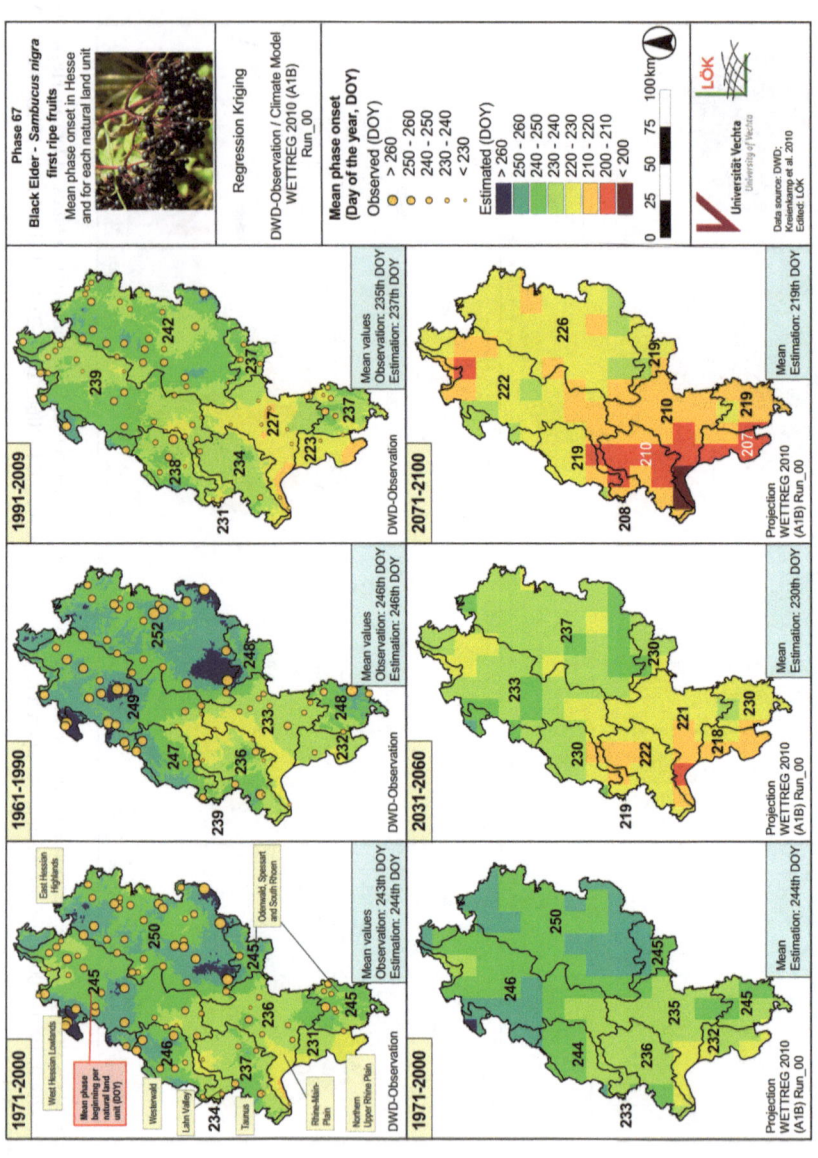

Fig. 3.15 Surface maps derived by Regression Kriging for the beginning of fruit ripening of *Sambucus nigra* (*black* elder) in Hesse based on observed (1961–1990, 1971–2000, 1991–2009) and projected air temperature grids (1971–2000, 2031–2060, 2071–2100) derived from WETTREG 2010 (run 0) climate model, emission scenario A1B

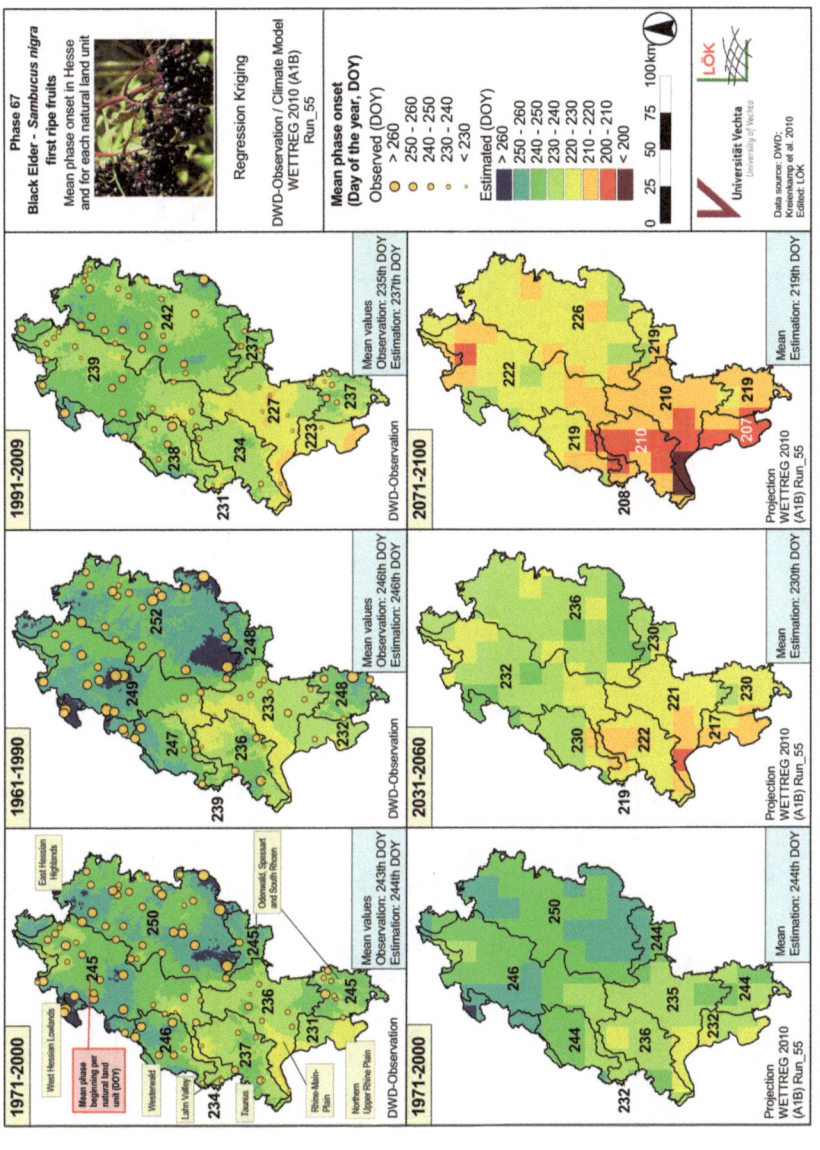

Fig. 3.16 Surface maps derived by Regression Kriging for the beginning of fruit ripening of *Sambucus nigra* (*black* elder) in Hesse based on observed (1961–1990, 1971–2000, 1991–2009) and projected air temperature grids (1971–2000, 2031–2060, 2071–2100) derived from WETTREG 2010 (run 5) climate model, emission scenario A1B

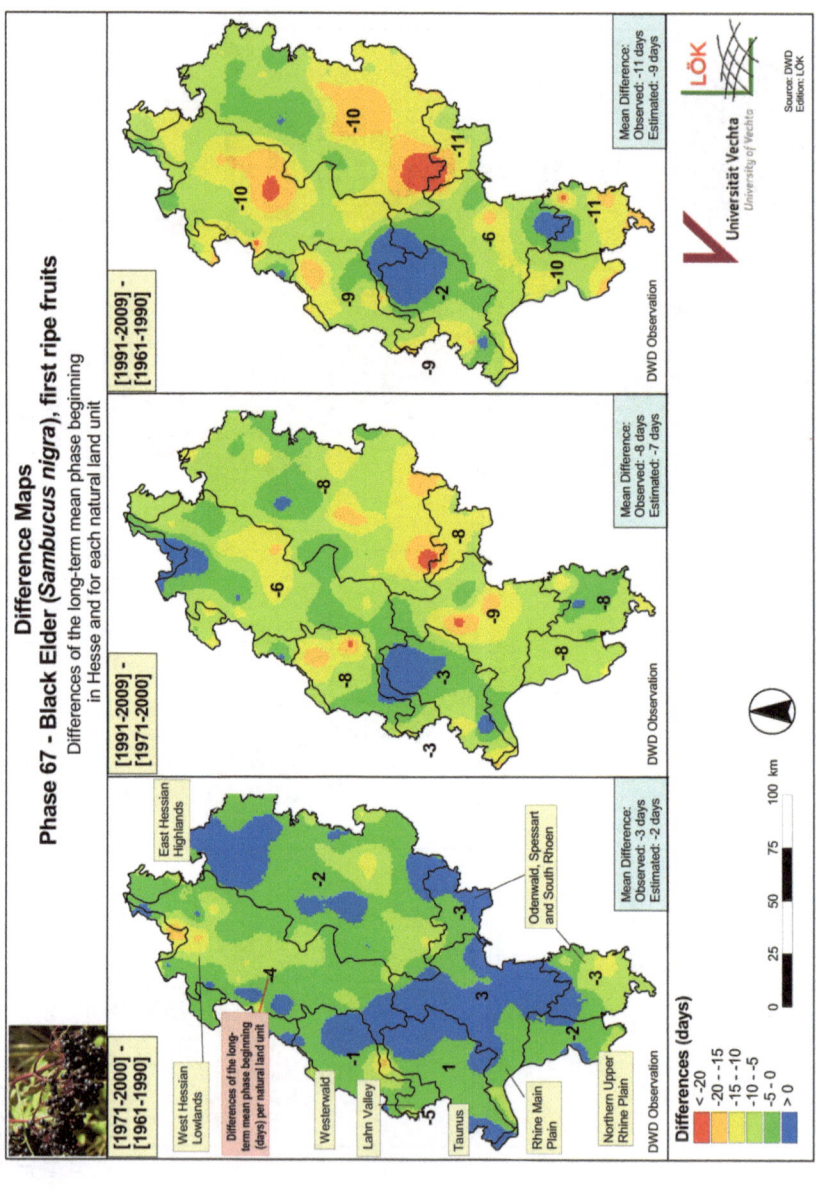

Fig. 3.17 Differences in the beginning of fruit ripening of *Sambucus nigra* (*black elder*) in Hesse in comparison of the periods 1971–2000 and 1961–1990 (*left*), 1991–2009 and 1971–2000 (*centre*), and 1991–2009 and 1961–1990 (*right*)

the beginning of the year was calculated for the Lahn Valley in the west of Hesse, for example. Some other land units even showed shifts towards the end of the year (Rhine-Main-Plain, Taunus). However, comparing the two younger periods, shifts towards the beginning of the year were calculated for all natural land units in Hesse and the maximum shift increased up to 9 days towards the beginning of the year (Rhine-Main-Plain). Locally, shifts of more than 20 days were calculated (red areas). Especially mountainous regions are affected by strong shifts towards the beginning of the year.

Considering the projected onset of the black elders' fruit ripeness of the periods 2031–2060 and 2071–2100, trends of the past should proceed in the future. However, the strength of the future shifts depended on the air temperature development as projected by the respective climate model (Sect. 3.2.2). The long-term mean shifts of phase 67 (fruit ripeness of *Sambucus nigra*) in Hesse between the periods 1961–1990 and 2071–2100 ranged between 25 days (REMO/UBA, Fig. 3.13) and 29 days (HADCM3/COSMO-CLM, Fig. 3.14). Accordingly, in average of the period 1961–1990 on 3 September (246th day of the year), the phase is expected to occur between 5 and 9 August (217th and 221st day of the year) in the long-term mean of the period 2071–2100 resulting in an overall shift of nearly 1 month. However, some areas in Hesse show even earlier onsets. For the Northern Upper Rhine Plain, the long-term mean phase onset was projected on 21 July (202nd day of the year) and for some areas in southern Hesse even before 19 July (less than 200 days after New Year). Considering some further individual characteristics of the four applied climate projections, there were distinct differences between the phenological development of Sambucus nigra comparing the climate reference periods 1971–2000, 2031–2060, and 2071–2100: Whereas ECHAM5/COSMO-CLM and REMO/UBA projected relatively late onsets for fruit ripening (phase 67) for the period 2031–2060 (233rd and 235th day of the year), HADCM3/COSMO-CLM and WETTREG2010 (run 0 and 5) showed stronger shifts between the earlier periods 1971–2000 and 2031–2060 since these two models already projected early phase onsets in 2031–2060 (229th and 230th day of the year) (Fig. 3.18).

Regarding the projected shifts of the respective phase onsets of all 23 phases analysed by Regression Kriging between the observed period 1961–1990 and the modelled period 2071–2100, the past phenological development will obviously continue until the end of the twenty first century (Table 3.3). With only few exceptions, the advance in phase onset should be at least twice as high as already observed comparing the climate periods 1991–2009 and 1961–1990. For many phases, they were even three times higher or more. For most of the phases, the onset is expected to shift towards the beginning of the year by 3 weeks and more, for instance, flowering of snowdrop (*Galanthus nivalis*, phase 2) or pears (*Pyrus communis*, phase 60) or fruit ripeness of the red currant (*Ribes rubrum*, phase 100). The strongest shifts were detected for phases occurring at the beginning of the phenological year (early and first spring) as well as for phases of fruit ripeness in mid- and late summer and early autumn (e.g., black elder). The beginning of hazel bloom (*Corylus avellana*, phase 1) is expected to be the most affected phase as shifts towards the beginning of the year of up to 40 days were estimated comparing the periods 1961–1990 and

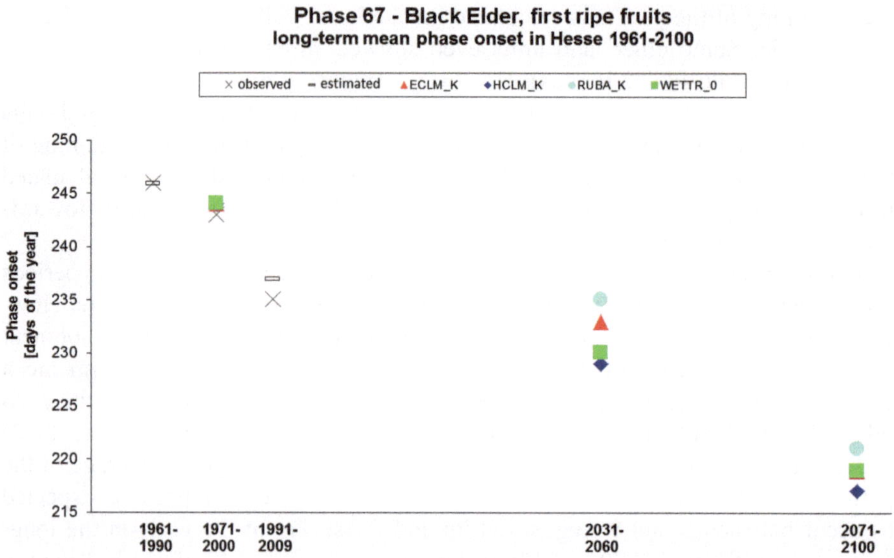

Phase 67 - Black Elder, first ripe fruits
long-term mean phase onset in Hesse 1961-2100

Fig. 3.18 Mean onset of the beginning of fruit ripeness of *black* elder calculated using observations (1961–2009) on phenology (observed=derived from local measurements provided by the German Weather Service, estimated=values derived from surface maps) and projections on phenological development based on modelled air temperatures (1971–2000, 2031–2060, 2071–2100) climate emission scenarios A1B

2071–2100. Fruit ripeness of the early ripening apple (*Malus domestica*, phase 109) is expected to begin in average between three and 4 weeks earlier till the end of the twenty-first century.

In summary, Table 3.3 illustrates to which extent the results calculated on the basis of the air temperature development as projected by the four climate models differ from each other. Phenological maps based on ECHAM5/COSMO-CLM and REMO/UBA were characterized by more conservative estimations as they project rather temperate temperature conditions and, thus, result in relatively late phase onsets. Considering the projection results of all 23 phenophases, the mean variation of projected phase onsets amounts to 6.5 days for the period 2031–2060 and to 4.7 days in the period 2071–2100 comparing the different climate projections.

For the three considered vine phases (Table 3.1, Sect. 3.2.1), a different approach was performed since only up to three observation sites were available for regression analysis. Consequently, surface estimations could not be realized. However, due to its economic relevance, estimations of the future phase onset for the examined vine phases were performed locally for the respective observations sites without calculating additional surface maps. Figure 3.19 shows the location of the three involved observation sites for vine flowering (phase 172) and grape gathering (phase 205). For phase 171 (sprouting), only two sites were involved (Unter-Hambach and Geisenheim) due to the lack of appropriate data.

Table 3.3 Differences in the phase onset of 23 plant phases comparing the long-term annual means of the observation periods 1961–1990 and 1991–2009 (*first row*) and the projected onset in the period 2071–2100 with the observation period 1961–1990 (*next 5 rows*)

Phase	1	2	6	52	7	115	62	13	15	19	18	123
Differences (days) between 1991–2009 and 1961–1990 (observations)												
	−14	−9	−11	−10	−8	−6	−10	−8	−9	−7	−10	−4
Differences (days) between 2071–2100 (projection) and 1961–1990 (observation)												
ECLM K	−30	−22	−23	−13	−16	−7	−16	−14	−15	−12	−17	−12
HCLM K	−32	−23	−26	−18	−19	−11	−20	−17	−18	−14	−19	−12
RUBA K	−33	−24	−25	−13	−17	−7	−15	−14	−15	−12	−16	−11
WETTR 00	−40	−28	−30	−18	−21	−11	−21	−18	−19	−15	−21	−13
WETTR 55	−40	−28	−30	−16	−20	−10	−19	−18	−17	−13	−19	−11
Phase	20	64	100	109	67	54	56	60	102	103	104	
Differences (days) between 1991–2009 and 1961–1990 (observations)												
	−7	−9	−8	−5	−11	−6	−7	−7	−12	−9	−10	
Differences (days) between 2071–2100 (projection) and 1961–1990 (observation)												
ECLM K	−15	−19	−25	−26	−27	−15	−15	−19	−26	−21	−24	
HCLM K	−15	−20	−25	−27	−29	−17	−18	−21	−25	−21	−24	
RUBA K	−14	−17	−22	−23	−25	−15	−14	−19	−23	−18	−20	
WETTR 00	−16	−22	−25	−27	−27	−20	−19	−25	−27	−21	−25	
WETTR 55	−15	−21	−26	−26	−27	−19	−17	−23	−26	−21	−25	

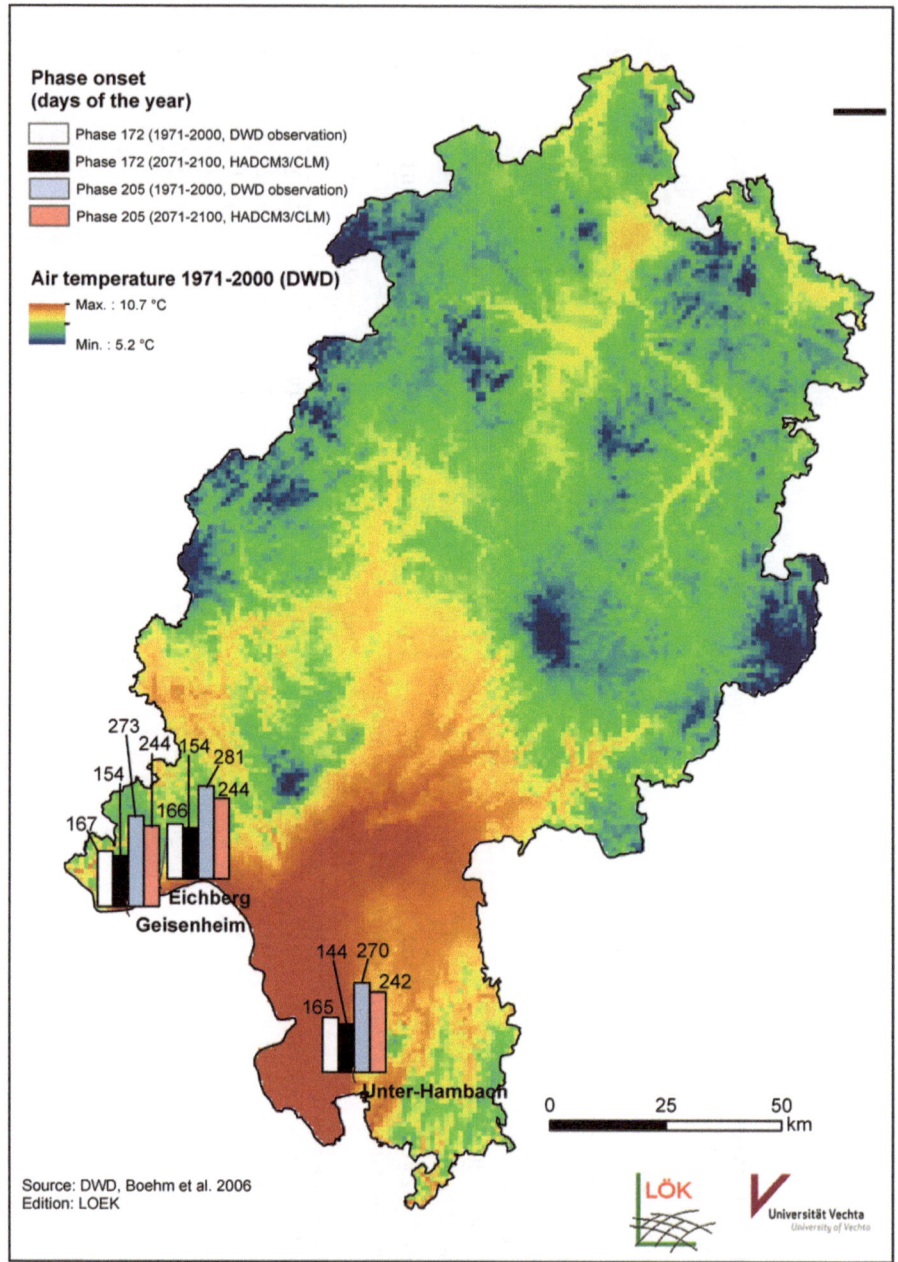

Fig. 3.19 Mean long-term annual onset of the flowering of wine (phase 172) and vintage (phase 205) at three observation sites in Hesse for the period 1971–2000 (observed data) and 2071–2100 (projected data, climate model HADCM3/CLM)

The charts on the respective phase onset calculated for each observation site (Fig. 3.19) show the differences between the observed long-term mean beginning of vine flowering (white bar) and vintage (light blue bar) in the reference period 1971–2000, compared to the respective onsets of the period 2071–2100 projected by example of the HADCM3/COSMO-CLM model (black bar = vine flowering, red bar = vintage).

In result, phase onset of vine at the three considered observation sites located in the south of Hesse showed a similar development compared to the other 23 phases examined by Regression Kriging: According to the respective climate model, vine flowering should occur in average between 2 and 3 weeks earlier in the period 2071–2100 compared to the period 1971–2000 whereas vintage is expected to take place 19 to 41 days earlier (Fig. 3.19). Considering the respective observation site and the particular climate model, grape harvest will take place between August 28 and September 8 in average of the period 2071–2100 instead of September 27 and October 8 as in the reference period 1971–2000.

When comparing the results with regard to the respective climate model one could see regional differences in phase onset even for the few observation sites: Referring to vine flowering (phase 172), at the locations "Geisenheim" and "Eichberg" in the Upper Rhine Main plain in the northwest a shift of phase onset of nearly 2 weeks was projected. In comparison, at the location "Unter-Hambach" in the Upper Rhine valley in the south of Hesse a shift of 3 weeks was projected. For grape harvest (phase 205) the reverse development was projected: Comparing the periods 1971–2000 and 2071–2100, phase onset should start 28 days earlier at the location "Unter-Hambach", but up to 37 days earlier when considering the two locations "Geisenberg" and "Eichberg".

3.5   Discussion and Conclusions

Especially for those regions in Hesse where no or only few phenological observations were available, the methodology introduced in this study is helpful to derive knowledge on the spatially differentiated plant development throughout the last 50 years and to assess the ongoing phenological development in the future. The quality of the surface estimations derived by Regression Kriging could be ensured by help of different statistical quality measures (e.g. root mean square error, RMSE): Considering all calculated surface maps, RMSE ranged between 0.08 and 1.5 days.

In comparison to the German-wide investigation (*case study 1*) presented in Sect. 2, the bivariate statistical analysis applied in *case study 2* for the federal state of Hesse was improved by using air temperature data of only those months that showed strong correlation between air temperature and phase onset instead of using annual mean air temperature data (Sec. 2.3). This enabled derivation of more powerful regression models indicated by higher correlation coefficients. Additionally, the calculation of residuals indicated the extent of influence of other drivers affecting phenological development. Even though the residuals could not be explained in terms of identifying latent factors, the spatial structure of the residuals was considered within the Regression Kriging approach used for mapping recent and potential future spatial and temporal trends of phenological developments. A refinement of the regression models presented in this study might be achieved by additional calculations for spatially and/or timely defined clusters in Hesse/Germany as demonstrated by Oteros et al. (2013).

In this investigation, only one emission scenario (A1B) was considered compared to *case study 1* (B1, A1B). To establish a more comprehensive range of projections of the impacts of climate change, the study might have used additional emission scenarios, for instance, more moderate estimations like B1 and B2 (IPCC 2007). However, within the framework of the research program INKLIM-A, the consortium, including experts for climate modelling, decided unanimously that it would make no sense to consider the B1 scenario since the underlying assumptions are meanwhile outdated. However, a wide range of temperature variation was considered by applying four climate models (instead of two for *case study 1*) reflecting both, more conservative and more extreme temperature developments.

The phenological maps derived in the HeKlimPh project allowed for developing appropriate adaption strategies to cope with this issue (Elith and Leathwick 2009; Schönrock et al. 2013). This may comprise, for instance, adjusting delineation, shape and geographical position of protected areas (Mawdsley et al. 2009; Milad et al. 2011). Furthermore, agricultural management has to consider selection of crops and cultivars under changed climatic conditions. Farmers have to cope with problems due to insect pests potentially raised by increased air temperatures (Bindi and Olesen 2011; van Vliet et al. 2013). Another threat is the increasing risk of frost damage due to the earlier occurrence of phenological events (Inouye 2000). A further problem to cope with will be the decrease of precipitation during hot and dry summers in Hesse which has to be dealt with by extensive irrigation measures during the vegetation period (Enke 2003, 2004). In contrast, there might be also beneficial effects of climate change in terms of a prolongation of the growing season (Henniges et al. 2005) leading eventually to increased yields in some regions. In the end, it might also be possible to grow new fruit varieties when the climatic conditions get appropriate (Bindi and Olesen 2011; Priess et al. 2005; Streitfert et al. 2005). Wine-growing regions might be affected by changing arrays of varieties due to climate change (Schultz et al. 2009). Thus, cultivation of varieties that previously only grew in southern regions might be possible in Hesse in the future (Stock et al. 2007).

The potential impact of the geographic distribution of the phenological observations across Hesse was not considered in the HeKlimPh-Project assuming that the observations were randomly distributed (van Vliet et al. 2013). This assumption could be proved in a succeeding study. Furthermore, the phenological monitoring should be complemented by a standardized metadata acquisition as done in the European moss survey (Pesch and Schröder 2006) to promote the interpretation of phenological data as Schönrock et al. (2013) showed in a local scaled pilot study. A further measure to improve validity of plant phenological monitoring may be the use of digital repeat photography (Crimmins and Crimmins 2008). Future approaches should also consider additional drivers for phenological development like supplementary climate elements (e.g., temporal and spatial distribution of precipitation), soil texture, and local land use. Phenological data supplied by the DWD feature some uncertainties caused by the survey methodology: Data are collected within a radius of up to 5 km around a fixed coordinate, representing the respective observation site. This means, accurate allocation of local environmental conditions at the respective observation sites is difficult. Metadata acquisition for each observed plant could help in this issue (Schönrock et al. 2013). Moreover, phenological data feature a temporal uncertainty as phenological observers conduct observations only 2 or 3 time per week. Hence, it is not given that the observed time of occurrence of a phase onset represents the real date. Additionally, each phenological observation depends on the experience and knowledge of the respective phenological observer (Schnelle 1955). Consequently, phenological observations may not completely fulfil the standards of empirical observations as described for instance by Schröder (1996) and Schröder et al. (1994) even if the observation guideline (DWD 1991) was followed. Further potential sources of error were mentioned by Schnelle (1955): Observation of invalid species or changing species within an observation site along the years of observation as well as genetic variation within the same species might lead to imprecise or misleading interpretation. In complementation to the plausibility checks processed by the DWD (2006), further statistical approaches to detect implausible values were introduced in this case study (Sect. 3.2.1).

The observed shift in phenological development was particularly determined by air temperature rise expected to continue or even speed up until the end of the twenty-first century. Accordingly, phenological records allow estimating future trends of plant phenological development and related ecological processes for agriculture, forestry, human health, and the global economy (Khanduri et al. 2008). As shown in this investigation, changes in plant phenology evidently reflect a warming trend. Thus, phenology is an appropriate bioindicator for mapping early signs of ecosystem transitions under climate change across areas of large spatial extend. Distinct effects on flora distribution might be species migration to higher latitudes and altitudes and eventually repressing domestic species. In the end, even the extinction of species may be possible due to changed habitat conditions (Theurillat and Guisan 2001). However, shifts through phenotypic plasticity occur prior to and more rapidly than the more profound changes in species distribution and genetics. Liang and Schwartz (2013) found that different plant genotypes require varying amounts of heat energy for starting annual growth and reproduction. It seemed that adaptation and other ecological and evolutionary processes along climatic gradients affected the timing of phenophases. Accordingly, earlier timing indicates higher efficiency, i.e., less heat energy needed to trigger phenophase transitions.

The phenological development of plants influences the mass (e.g., carbon dioxide and water) and energy cycle of the biosphere. For the beginning of leaf formation of wheat, barley and rapeseed in Germany Ma et al. (2012) calculated an advance of 1.6–3.4 days per decade during 1961–2000. Shiferaw et al. (2013) reported that due to temperature increase since 1980 the worldwide wheat yields decreased by 5.5 % without considering the effect of increasing CO_2 levels, and by 2.5 % when considering C-fertilization. Up to 2050, wheat yield levels are expected to decrease by 5–9 % for ombrogenous cultivation systems emphasizing the importance of plant phenological observations not only for nature protection but also for agricultural management issues.

References

Bindi M, Olesen JE (2011) The responses of agriculture in Europe to climate change. Reg Environ Change 11:151–158

Böhm U, Kücken M, Ahrens W, Block A, Hauffe D, Keuler K, Rockel B, Will A (2006) CLM-the climate version of LM: brief description and long-term applications. COSMO Newsl 6:225–235

Brown I (2012) Influence of seasonal weather and climate variability on crop yields in Scotland. Int J Biometeorol 57:605–614

Brown DG, Aspinall T, Bennett DA (2006) Landscape models and explanation in landscape ecology—a space for generative landscape science? Prof Geograph 58:369–382

Crimmins MA, Crimmins TM (2008) Monitoring plant phenology using digital repeat photography. Environ Manage 41:949–958

Dale MRT, Fortin MJ (2009) Spatial autocorrelation and statistical tests: some solutions. J Agr Biol Environ Stat 14:188–206

Dutilleul P (1993) Modifying the t-test for assessing the correlation between two spatial processes. Biometrics 49:305–314

DWD (Deutscher Wetterdienst) (1991) Anleitung für die phänologischen Beobachter des Deutschen Wetterdienstes. Deutscher Wetterdienst, Offenbach

DWD (Deutscher Wetterdienst) (2006) Phänologie-Journal. Mitteilungen für die phänologischen Beobachter des Deutschen Wetterdienstes 27. Offenbach

Elith J, Leathwick J (2009) Species distribution models: ecological explanation and prediction across space and time. Annu Rev Ecol Evol 40:677–697

Enke W (2003) Anwendung eines Statistischen Regionalisierungsmodells auf das Szenario B2 des ECHAM4 OPYC3 Klima-Simulationslaufes bis 2050 zur Abschätzung Regionaler Klimaänderungen für das Bundesland Hessen. Hessisches Landesamt für Umwelt und Geologie, Wiesbaden

Enke W (2004) Erweiterung des Simulationszeitraumes der Wetterlagenbasierten Regionalisierungsmethode auf der Basis des ECHAM4-OPYC3 Laufes für die Dekaden 2011/2020 und 2051/2100, Szenario B2. Hessisches Landesamt für Umwelt und Geologie, Wiesbaden

Fortin JM, Payette S (2001) How to test the significance of the relation between spatially autocorrelated data at the landscape scale: a case study using fire and forest maps. Ecoscience 9:213–218

Hagl S (2008) Schnelleinstieg Statistik: Daten Erheben, Analysieren, Präsentieren. Haufe-Lexware, München

Hengl T, Heuvelink GBM, Rossiter DG (2007) About regression-kriging: from equations to case studies. Comput Geosci 33:1301–1315

Henniges Y, Danzeisen H, Zimmermann RD (2005) Regionale Klimatrends mit Hilfe der phänologischen Uhr, dargestellt am Beispiel Rheinland-Pfalz. Umweltwiss Schadst Forsch 17(1):28–34

Holopainen J, Helama S, Lappalainen H, Gregow H (2013) Plant phenological records in northern Finland since the 18th century as retrieved from databases, archives and diaries for biometeorological research. Int J Biometeorol 57:423–435

Inouye DW (2000) The ecological and evolutionary significance of frost in the context of climate change. Ecol Lett 3:457–463

IPCC (Intergovernmental Panel of Climate Change) (2007) Climate change 2007. Synthesis report. Geneva, 52 p

Jacob D, Göttel H, Kotlarski S, Lorenz P, Sieck K (2008) Klimaauswirkungen und Anpassung in Deutschland-Phase 1: Erstellung regionaler Klimaszenarien für Deutschland. Forschungsbericht 204 41 138, UBA-FB 000969. Clim Change 11:1862–4359. http://www.umweltbundesamt.de/sites/default/files/medien/publikation/long/3513.pdf. Accessed 14 July 2014

Keuler K, Lautenschlager M (2006) Climate Simulations with CLM. Climate of the 20th Century run No. 1, 1960-2000, Data Stream 2 und Scenario A1B run No. 1, 2001-2100, European Region,

MPI-M/MaD. Max-Planck Institut für Meteorologie, Hamburg. http://cera-www.dkrz.de/WDCC/ui/BrowseExperiments.jsp?proj=CLM_regional_climate_model_runs. Accessed 14 July 2014

Khanduri VP, Sharma CM, Singh SP (2008) The effects of climate change on plant phenology. Environmentalist 28:143–147

Kreienkamp F, Spekat A, Enke W (2010) Ergebnisse eines Regionalen Szenarienlaufs für Deutschland mit dem Statistischen Modell WETTREG2010. Report. CEC, Potsdam

Legendre P (1993) Spatial autocorrelation: trouble or new paradigm? Ecology 74:1659–1673

Li Z, Yang P, Tang H, Wu W, Yin H, Liu Z, Zhang L (2014) Response of maize phenology to climate warming in Northeast China between 1990 and 2012. Reg Environ Change 14:39–48

Liang L, Schwartz MD (2013) Testing a growth efficiency hypothesis with continental-scale phenological variations of common and cloned plants. Int J Biometeorol. http://link.springer.com/article/10.1007%2Fs00484-013-0691-6

Ma S, Churkina G, Trusilova K (2012) Investigating the impact of climate change on crop phenological events in Europe with a phenology model. Int J Biometeorol 56:749–763

Mawdsley JR, O'Malley R, Ojima DS (2009) A review of climate-change adaptation strategies for wildlife management and biodiversity conservation. Conserv Biol 23(5):1080–1089

Meynen E, Schmithüsen J, Gellert J, Neef E, Müller-Miny H, Schultze JH (1953–1962) Handbuch der naturräumlichen Gliederung Deutschlands, 2 Bd. Bad Godesberg

Milad M, Schaich H, Bürgi M, Konold W (2011) Climate change and nature conservation in Central European forests: a review of consequences, concepts and challenges. Forest Ecol Manage 261:829–843

Odeh IOA, McBratney AB, Chittleborough DJ (1995) Further results on prediction of soil properties from terrain attributes: heterotopic cokriging and regression-kriging. Geoderma 67:215–226

Oteros J, García-Mozo H, Hervás-Martínez C, Galán C (2013) Year clustering analysis for modelling olive flowering phenology. Int J Biometeorol 57:545–555

Parmesan C (2007) Influences of species, latitudes and methodologies on estimates of phenological response to global warming. Glob Change Biol 13:1860–1872

Pesch R, Schröder W (2006) Assessment of metal accumulation in mosses by combining metadata, statistics and GIS. Nova Hedwigia 82(3–4):447–466

Priess JA, Heistermann M, Schaldach R, Onigkeit J, Mimler M, Trinks D, Alcamo J (2005) Klimawandel und Landwirtschaft in Hessen: Mögliche Auswirkungen des Klimawandels auf Landwirtschaftliche Erträge. Abschlussbericht für den Bereich Landwirtschaft, InKlim 2012-Integriertes Klimaschutzprogramm Baustein II: Klimawandel und Klimafolgen in Hessen. http://klimawandel.hlug.de/fileadmin/dokumente/klima/inklim/endberichte/landwirtschaft.pdf. Accessed April 2014

Romić MT, Hengl D, Romić D, Husnjak S (2007) Representing soil pollution by heavy metals using continuous limitation scores. Comput Geosci 33:1316–1326

Rosenzweig C, Karoly D, Vicarelli M, Neofotis P, Wu Q, Casassa G, Menzel A, Root TL, Estrella N, Seguin B, Tryjanowski P, Liu C, Rawlins S, Imeson A (2008) Attributing physical and biological impacts to anthropogenic climate change. Nature 453:353–357

Schnelle F (1955) Pflanzen-Phänologie. Geest & Portig, Leipzig

Schönrock S, Schmidt G, Schröder W (2013) Klimabiomonitoring. Untersuchung der Pflanzenphänologie auf lokaler Ebene und ihr Vergleich mit regionalen und nationalen Daten. Natur Landschaft 88:14–21

Schröder W (1996) Einsatz von Biosphärenreservaten für Integrative Umweltbeobachtung und -bewertung sowie Naturschutz. Beiträge der Akademie für Natur- und Umweltschutz Baden-Württemberg 23:143–167

Schröder W, Vetter L, Fränzle O (eds) (1994) Neuere statistische Verfahren und Modellbildung in der Geoökologie. Vieweg, Braunschweig

Schultz H R, Hofmann M, Jones G (2009) Weinanbau im Klimawandel: Regionen im Umbruch. Klimastatusbericht des DWD, 12–20

Shiferaw B, Smale M, Braun H-J, Duveiller E, Reynolds M, Muricho G (2013) Crops that feed the world 10. Past successes and future challenges to the role played by wheat in global food security. Food Sec 5:291–317

Stock M, Badeck F, Gerstengarbe F W, Hoppmann D, Kartschall T, Österle H, Werner P C, Wodninski M (2007) Perspektiven der Klimaänderung bis 2050 für den Weinbau in Deutschland (Klima 2050). Schlussbericht zum FDW-Vorhaben: Klima 2050. Potsdam Institute for Climate Impact Research (PIK). PIK report 106, Potsdam

Streitfert A, Grünhage L, Jäger H-J (2005) Klimawandel und Pflanzenphänologie in Hessen. Institut für Pflanzenökologie, Justus-Liebig-Universität Giessen

Theurillat JP, Guisan A (2001) Potential impact of climate change on vegetation in the European Alps: a review. Clim Change 50:77–109

van Vliet AJH, Bron WA, Mulder S, van der Slikke W, Ode B (2013) Observed climate-induced changes in plant phenology in the Netherlands. Reg Environ Change 14(3):997–1008. doi:10.1007/s10113-013-0493-8

Zirlewagen D, Raben G, Weise M (2007) Zoning of forest health conditions based on a set of soil, topographic and vegetation parameters. Forest Ecol Manage 248:43–55